LOTTERY MAESTRO

Jackpot Edition

Eze Ugbor

Preface

The making of this book took over 30 years. This is the most comprehensive lottery book of its kind. I painstakingly ran through all the calculations. It works everywhere the lottery is played. Back test the book as far as possible. "Lottery Maestro" gives you precision forecasting.

I would like to first give thanks to The Almighty God for using me as the medium to accomplish this book that a lot of people believed was impossible. I would like to thank my family and friends for being there with me through this difficult process of countless equations and predictions. I would like to thank you as well for giving me the reason to continue in my pursuit.

Whatever you do, make sure you read the "Lotto" section and learn how one key number actually produced all winning Powerball numbers and how you can apply the same methodology to achieve similar results. Make sure to study the "Chapter Supplement" where you will master the "Lottery Maestro" signature: the only lottery system that has never, and will never, fail.

"Lottery Maestro" will put you outside the realms.

Thank you,

Eze Ugbor

Table of Contents

Chapter 1

Number Difference Formula

Let's start by lining up a simple set of Pick 3 numbers to create a trend which will be expanded upon later in this book to show you different ways of winning more money. We are going to create winning Pick 3 numbers from the Maryland lottery, beginning April 1st 2019. Please bear in mind that it doesn't matter which date or state lottery this book starts from.

All trends exhibit similar behavior regardless of the lottery market. You will achieve the same goal no matter which corner of the globe you're betting from.

Sample Table 1

0	1	2
3	4	5
6	7	8
8	9	0
1	2	3

The above table is very easy to form once you understand the method being used. I placed the numbers in numeric order with a little twist towards the end where I introduced the number 8 twice. You should be able to figure out how to recreate the above group with careful examination. Also note that I could have repeated any of the numbers, like I did with the number 8, or decided not to do that. It really doesn't matter because you will have mastered the art of overcoming any lottery betting obstacle by the time you reach the end of this book.

Now let us examine the above winning Pick 3 numbers against the Maryland lottery winning Pick 3 numbers beginning from April 1st 2019.

The assumption is that the Maryland lottery results for April 2019 are the only information available to you. At this point you are studying the winning numbers from the month of April to prepare you for more winners down the line. That study includes recreating winning numbers based on Sample Table 1.

What you will notice right away is that winning number 678 was played as 786, which would pay $40 on 50 cents box. You would have likely played it as 678, which was the information at hand, but you would likely not win because that was the beginning of the month. At the start of the month of April, you most likely will not have enough information to capture the winning numbers yet. You need to take time to study the trend, and recreate the trend based on what is playing to be able to capture winners.

The next winning number that played among the group is 231, which played almost two weeks later on 4/13/2019. That also would pay out $40 on 50 cents box. The rest of the group didn't win for the entire month. Even if you had won both it wouldn't be something to write home about. After all, that would only be winning $80 for the entire month.

When I explained this to a friend, he responded with the truth: "it would not be encouraging by any measure".

That's the elephant in the room. With all that said, we must tackle the crucial, unavoidable question.

What do I need to do to capture more winnings?

The answer lies in being able to make the necessary adjustments based on what trend is playing, mastering how to expose the hidden numbers for the first time ever, and of course employing the ultimate formulas; Shadow, Counterpart, and of course Numbers.

Put on your seatbelt and come along for the ride.

Let's bring the Sample Table down again before we continue, just to remind ourselves what numbers we're dealing with.

Sample Table 1

0	1	2
3	4	5
6	7	8
8	9	0
1	2	3

Combing for more information within the numbers will expose (for the readers who don't know yet) what we call the Number Difference formula, in lottery jargon. Those will give you easy winnings once you understand them.

One example will be the winning number 786 that played on April 1st 2019. That is simply a variation of the sequential number 678, which is in our Sample Table 1. In lottery jargon this is called Difference 1. Anytime a Difference 1 is played, one or more of the numbers in the group is bound to drop in similar order. In other words, 678 will produce similar winning numbers, such as 456, 567, 678, 789, 890, and so on. You will notice that the set of Pick 3 numbers have at least one winning number from the group that played in Difference 1 formation. That winning number played on 4/9/19 as 546. For the doubters, check any lottery result from any country in the world to confirm the Difference 1 formation. It works the same for Pick 4 numbers.

On close examination you will notice another Difference 1 that played as 231 on 4/13/19, followed by 901 on 4/17/19. In case you are wondering how to know when the next one is playing, the winning number 546 set the trend for the rest of the group. The 546, 231 and 901 played exactly 5 days apart from each other. I assure you, this is far from a coincidence.

I am not in a position to tell you exactly when the next one is going to drop, but be on the lookout now that you know what to look for. The same method goes beyond Difference 1. One example of Difference 2 would be winning number 135 on 4/11/19 and 579 on 4/20/19.

There's a bunch of them that will give you more winnings, including Difference 3. I will lay out some of the groups to show you the efficacy of this simple method. You should strive to comb out the rest, because that will mean more winnings in your pocket. The method is simple yet solid.

Some of the groups and the time line they HIT.

789 to 549 played 9 days apart

280 to 135 played 4 days apart

135 to 417 played 5 days apart

546 to 231 played 5 days apart

231 to 901 played 5 days apart

901 to 579 played 4 days apart

579 to 147 played 4 days apart

417 to 147 played 9 days apart

You will notice that the beginning and end of the trend played 9 days apart on each case.

The rest of the groups played 3 among them in 5 day intervals and the other 3 played 4 days apart. There is no better way to exemplify the precision here.

At this point, if people still want to call the correlation between these winning numbers a coincidence, I doubt there will be any convincing them. Group the winning numbers in vertical order and take a look at past results against them. A careful study will clue you in to the relationship between the numbers, and thus help you know what to play, that will give you a parlay win at the very least. Remember that preservation of capital is the key. An example will be taking a look towards the last three winners, 579, 147 and 147. Placed in vertical order, they become 977, 744 and 511. Maryland lottery played recently 778, 744 and 611 on 4/11/20 through 4/15/20. You can see how close the two groups are. By the way, Maryland played the 3 winning Pick 3 numbers 2 days apart. That should be enough to show you the power of the Difference Number method.

In the interest of time and efficiency, I won't go into a complete line by line breakdown of the entire group. If you analyze the numbers for yourself, you will see this trend reflected in the winning numbers that were played in the Maryland lottery. I have to bring in the other methods that will give you a whole lot more winning numbers.

Readers have an insatiable appetite for winning, and because of that we must explore further. I'm sure you all have no issue with potentially getting twice as many winning numbers, or more.

You will notice that in the month of April 2019 there are numerous double winning numbers like 411, 994 and so on. You actually win more money on those, which necessitates that we work on deconstructing those in order to understand them. There's also the Pick 4 which we have not yet touched either. Do not forget the 6/49 and other lotto games. If you are a student of lottery, you've probably heard people talking about the balanced game, balanced wheel and balanced numbers.

The real question is, balanced from what? If the numbers being used are skewed you can never come close to the so-called "balance," even if the people pitching that are screaming from the rooftops, claiming that their books and methods are number one. It is not enough. There is a big difference between throwing out a bunch of numbers and wheeling them and actually working out the numbers based on the game you are betting on. If you construct a lottery table like the ones I explained how to make in one of my earlier books, "Lottery Icon" you can use the tables in that book to bet on any lotto game in the world. If you are talking about getting to something close to the word "balance" when it comes to the lottery, there is only one number. That number plays a central role in any lotto game you are betting on. It took years to zero in on that number. You are going to see that number and also master how to discover it for yourself by the end of this book.

The rest of the book will bring in the other winning formulas as well as touch the Counterpart, Shadow and Number methods. These will show you how to expose the hidden trends and winning numbers that you would not otherwise notice.

Chapter 2

Winning Numbers Exposed

As you can see, we have produced several winning Pick 3 numbers for the month of April 2019 via simple observation using the Difference Number methods. That includes Difference 1, Difference 2 and more. It's almost like picking winning Pick 3 numbers without lifting a finger. I'm sure you're wondering whether more winning numbers can be produced from the Sample Table. Luckily for you, I'm about to show you exactly how that can be done.

We'll be using the same set of numbers, rearranging the group with two things to keep in mind; one of which is introducing the winning Pick 3 number 276 into the equation, and the other is knowing that some of the winning numbers are actually hidden on the sides.

You are going to learn how to achieve these by doing the exercises to create Sample Table 2, after which we will show you how to win the Pick 4 numbers, bypassing the limits of the so-called "1-in-10,000" odds.

Now let's bring down the Sample Table 1, expand on it, and begin to crank out endless winning numbers.

Sample Table 1

0	1	2
3	4	5
6	7	8
8	9	0
1	2	3

The winning number exercise from Chapter 1 focused largely on the winning number 786 that played on 4/1/19. We are now going to create Sample Table 2. The winning number 276 will be central since it is an accomplice to 786, as both winning numbers played the same day of 4/1/2019.

Exercise 1

0

1

2

Workout 2

0

1

2

3 4 5

Workout 3

0

1

6 2 7

3 4 5

Workout 4

8 0

9 1

6 2 7

3 4 5

Workout 5

8 0

9 1 0

6 2 7

3 4 5

Workout 6

8	0	
9	1	0
6	2	7
3	4	5
1	2	3

Workout 7

8	0	**8**
9	1	0
6	2	7
3	4	5
1	2	3

For the readers who are wondering why the last number placement was employed to create 808, remember that Sample Table 1 had the number 8 twice in the group which means that Sample Table 2 must have the same repetition of numbers, because we used the exact same set of numbers in Sample Table 1 to create Sample Table 2.

Sample Table 2

8	0	**8**
9	1	0
6	2	7
3	4	5
1	2	3

You will notice right away that Sample Table 2 produced winning numbers 123, 627, 910 and 808. You can clearly see how many winning numbers we have produced from 2 simple tables in one month. There's actually a lot more winnings to be had in the above 2 tables by the time we introduce the Shadow and Counterpart methods.

As I've stated previously, there are 3 key methods in lottery betting called Number, Shadow and Counterpart. The two simple tables from earlier made use of only the Number method. We are going to introduce the other two, Shadow and Counterpart. If and when necessary we will use the acronym N, S & C to represent Numbers, Shadow and Counterpart.

We will put down the table for the three related methods that will show you how the numbers rotate. Most readers of our books will already be familiar with N, S & C. If you are new to these methods, make every effort to master them, because you need to understand them in order to produce a greater amount of winnings. The implementation of N, S, & C could make the difference between winning 5 times in a month and winning twenty times or more over the same period.

We are now going to use the Sample Table 2 group and show you how to produce winning Pick 4 numbers. At this point please remember that all of the winning numbers are related. The same tables can be used to win lotto games including Powerball, Mega Millions, Euro Millions, Lotto Max, and every other lotto game in the world.

Let's bring down Sample Table 2 and use it to recreate and examine some winning Pick 4 numbers from the month of April 2019. If we can produce several winning Pick 4 numbers in the same month from that simple table it should call into question the notion of there being 1:10,000 odds of winning Pick 4.

Granted, the unlikely odds might become true when you are limited to the elementary knowledge of the numbers which most people have; that is, simply looking at them through the lens of numeric order.

The numeric order is, by nature, nothing short of brainwashing. If you are constantly reinforced to count in order (1, 2, 3, 4, and so on) throughout your life then that becomes the only way you think to interpret numbers. Your vision will always be limited to the space between you and a large pillar in front of you, blocking your view. The only way you can see the larger space that exists beyond the pillar is to open your mind, and muster the courage to step away from it and see the bigger picture. When you master this retraining of the mind, you will quickly realize that there is so much more to be seen than once thought.

That metaphorical space represents winning more lottery numbers consistently. That space represents reprogramming your way of thinking by making the effort to master the Number, Shadow and Counterpart methods. Stepping away from the pillar and revealing that "space" is the equivalent of revealing the hidden relationships between numbers that you otherwise would not even know existed, let alone be able to understand and use them to your advantage.

I encourage you to go and study N, S & C through my other book, "Lottery Icon", which delves into much greater detail regarding those than I will in this book. Before people accuse us of using this as an excuse to promote that book, I encourage you to go to your local library to see if you can borrow it free of charge. The point here is not to advertise as many books as possible, but to help you fully understand my methods, so that you can increase your winnings. I'm not mandating that you read "Lottery Icon"; if you don't want to master the methods, then that's up to you.

With that out of the way, allow me to give you the N, S & C table before we continue working on the Pick 4.

Number, Shadow & Counterpart Table

#	0	1	2	3	4	5	6	7	8	9
S	3	4	8	0	1	7	9	5	2	6
C	8	9	3	5	6	2	4	0	7	1
S	2	6	0	7	9	8	1	3	5	4
C	7	1	5	2	4	3	6	8	0	9
S	5		7	8		0		2	3	
C	0		2	3		5		7	8	

The Number, Shadow and Counterpart table above is what you need to stay ahead of the curve. That is the table that will generate you more winning numbers. It might look daunting at first glance, but it's actually relatively easy to master. For the sake of simplicity, I will be referring to each column of numbers as a rotation. You can begin by studying the rotation under columns 1 and 2, or the one under columns 6 and 7 because they cover all of the digits, from 0 to 9.

You should try and memorize the rotations until you can recite them off of the top of your head. If you add up the amount of numbers in all of the rotations you will notice that there are 52 numbers in the entire set. It's interesting to note that there are 52 weeks in 1 year. Maybe there's a correlation there, but for now let's just concentrate on the issue at hand, which is winning the lottery consistently.

If you find it challenging to memorize the whole table, focus on mastering the number rotations in columns 1 & 2 because they more or less represent the entire table. Spend all the necessary time you need to master those, because once we go into Pick 4 and other lotto games it will become highly important to have that knowledge. We are not going to spend much time on that beyond this point. Our goal is to create a short, concise book with all the information you need to win every lotto game in front of you.

Now, let's bring down Sample Table 2 and show you how to beat the common Pick 4 axiom of 1:10,000 odds.

Sample Table 2

8	0	8
9	1	0
6	2	7
3	4	5
1	2	3

P4 Exercise 1

8	0	8	4
9	1	0	3
6	2	7	
6	2	6	1
3	4	5	2
1	2	3	0

("7" is shown as a decoy to produce the number 1, thereby making it 6261)

P4 Exercise 2

8	0	8	4
9	1	0	3
6	2	6	1
3	4	5	2
1	2	3	0

What you will notice is that the P4 exercise is derived largely from Sample Table 2 with the exception of the number 2, which came from outside to join the 345, making it 3452. Essentially, the pertinent information here is that the two winning numbers in the group that got altered were the 627 and 345.

The reason the number 2 was brought to accompany 345 in the Pick 4 exercise is to complete the order. You can see that the Pick 3 group changed into Pick 4 with numbers

0, 1, 2, 3 and 4. The winning Pick 3 number 627 became 6261 by subtracting 1 from 7 and carrying that over into the fourth digit position. If that number was left as 627 and 1 was added onto the end of that it would create 6271, you will in essence have a Number formation without the Counterpart. The Counterpart of 6 being 1 and the counterpart of 7 being 2. The equation will not balance because you have the original number as 627 which is Number and Counterpart (7&2). In order for the trend to continue the new group must have Number and Counterpart like the winning number 627, which is why the Pick 4 number became 6261 thereby accomplishing the goal of having a Number and Counterpart (6, 2, and 6) with an imported number (1).

One important thing worth noting is that a number in the group played twice (the number 6) in the process of creating the Pick 4 with the Counterpart 1, but the number 7 was also changed. 7 was split into 6 and 1 in order to maintain the original sum.

This means that the trend altered, and because of that the winning number 345 must also alter in line with the new formation. The 345 has already been established as now being 3452, but must change to be in line with the trend set by 6261. It now has to have a double number in the group. Let's continue forward to understand how that can be executed.

P4 Exercise 2

8	0	8	4
9	1	0	3
6	2	6	1
3	4	5	2
1	2	3	0

P4 Workout 3

6	2	7
3	4	5

The shadow of 7 is 5, and that is the first thing in common between the 627 and 345. The 627 now morphed into the Pick 4 number 6261. In order for the 345 (or 3452 as the case may be) to continue to stick to the established relationship, the next related number it can cling to is the number 2 from 627. It has to for more than one reason; the shadow of 7 (taken from 627), which is 5, will no longer work because that 7 has now been split into 6 & 1. The number 5 has no relationship with those, so the only choice left is to attach as the Counterpart of 2 from 627. It will now become 7 because the number 7 cannot just disappear. The trend is shifting in a hidden format that is only exposed through the application of Shadow and Counterpart.

The sum of the digits of 3452 is 14.

How do you bring in the 7 as a Counterpart and retain the same pattern? The group must include a double number for the simple reason that 627 was travelling with 345 all along. The sum of 627 is 15. Even when converted to the 6261 we see in Pick 4 the sum still remains as 15.

As long as the sum of 627 or 6261 is 15, the sum of 3452 must be 14 for the trend to continue.

You have to bring down the number 7 to include the only related number from 6261, and also change the remaining group from 3452 to create a double number. Since we're trying to maintain a sum of 14, the only double number that can be used while also including the number 7 is 3 & 3. Once you do that the sum of the digits will be 6 plus the 7 resulting in 13, and what's left will be 1. The sum total, therefore, is back to our target of 14.

Once this is done the group is now 7313 which completes the trend.

Now let's bring down the entire group and see if the effort is worth celebrating.

Sample Table 4

8	0	8	4
9	1	0	3
6	2	6	1
7	3	1	3
1	2	3	0

Now let's take a look and see if some of the winning Pick 4 numbers in the trend played during the month of April 2019.

Pro Table 1

8084 played on 4/2/19

7313 played on 4/5/19

1930 played on 4/6/19

1230 played on 4/12/19

6261 played on 4/23/19

Since you now know that you can hone in on 5 winning Pick 4 numbers in just one month from a simple Pick 3 table, clearly those 1:10,000 odds are nothing to concern yourself with.

However, this won't end with you magically winning that many Pick 4 numbers in just one month. I am not going to sit here and tell you that it's a cakewalk, but trust me when I say that continued effort will be worthwhile. We are talking about a game that pays out $5000 on a straight $1 win. So study the group as much as you need to because that will make it easier for you down the line.

As you can see, the above winning Pick 4 numbers would not be possible without the application of Shadow and Counterpart.

The above group is what we call a perfect winning Pick 4 group. It shows you the winning trend, and on a larger scale, you can use it to work out more winning numbers in the months to come. Remember that the group exhibits similar behavior because the trend will always be in place. The numbers play on different dates to try and obscure the trends, but with careful study and the application of N, S & C you will be picking out the winning numbers.

We've made progress, but there are more tasks ahead. If you are still unsure of how we got to certain conclusions, go back in the chapter and study the material further.

With all that being said, once you fully understand the material thus far.

Chapter 3

Winning Across the Board

In the last chapter you learnt how you can actually take Pick 3 numbers in a trend and turn them into winning pick 4 numbers. The reverse can also be the case where you turn Pick 4 into winning Pick 3 numbers. The simple reason for that is because all numbers are the same in the world of lotto games. The only thing you need to bring in the midst is your know how of N, S & C to expose the numbers you won't otherwise recognize.

You can also work out winning numbers using results from one state or country to another. We are going to demonstrate this using California Pick 4 winning numbers to produce Maryland winning Pick 3 numbers with two different time frames.

Let's begin this exercise by putting simple trends that everybody will understand.

Simple Trend

0	1	2	3
4	5	6	7
8	9	0	1
2	3	4	4

A closer look will show you that we repeated the number 4. Please note that it could be any number among the groups or none as you deem fit. It doesn't matter because N, S & C will put the trend in place regardless.

California Pick 4 played 9587 on 11/2/19. Our simple trend above have the numbers 8 and 9 among the group. We are trying to recreate the winning number 9587. The numbers in the group can be changed between vertical and horizontal order as necessary because they are not set on stone. The more important point is that you keep the same set of numbers in the group to maintain the trend.

We will start the process by putting down the 8901 in the form of:

1st Step

1

0

9

8

We couldn't leave it in the original order (8901) because the fourth digit number (1) cannot be converted to produce the 5 or 7 that will be needed to create 9587. The only solution will be to change the order and shift the rest of the group to produce 9587.

2nd Step

1	7
0	6
9	5
8	4

As you can see we now have 9 & 5 in the quest to create 9587. The next steps must shift the group in the order that will produce 8 & 7 to complement the 9 & 5 from 2nd Step above.

3rd Step

1	7	4
0	6	4
9	5	3
8	4	2

We now have 953 because the number 3 can be converted into Shadow and Counterpart to produce 7 or 8 as the case may be.

1	7	4	0
0	6	4	1
9	5	3	2
8	4	2	3

The readers that might find 4th Step a bit confusing should note that they are the same numbers from the simple trend group,

0	1	2	3
4	5	6	7
8	9	0	1
1	2	4	4

They are the same group switched from horizontal to vertical in the necessary order that will produce the winning California Pick 4 number 9587 that played on 11/2/19.

We already have 9 & 5 in the 4th Step group. All we need to do at this point is to convert the 3rd and 4th digits using Shadow and Counterpart to create 7 & 8.

4th Step

1	7	4	0
0	6	4	1
9	5	3	2
8	4	2	3

At this juncture that goal could be easily achieved by changing the last two digits into Counterpart. Remember that you produce the Counterpart by just adding 5 to the number. You have to change the last two digits for the trend to remain in place.

For instance the Counterpart of 4 is 9 and the Counterpart of zero (0) is 5.

The previous group after Counterpart application will now become:

Master Table

1	7	9	5
0	6	9	6
9	5	8	7
8	4	7	8

We have now recreated the California Pick 4 winning number 9587 that dropped on 11/2/19.

California lottery played 0960 on 11/3/19 and 3261 on 11/4/19.

We are trying to use the simple trend and recreate the California Pick 4 numbers that played on 2, 3 and 4th (9587, 0960 and 3261) of November 2019.

Master Table 1

We are going to create 0960 from Master Table above which will involve converting the Pick 4 in the first row 1795 into Counterpart which would become 6240 as well doing the same for the Pick 4 numbers in the 1st column 1098 into 3456 and that of the 4th digit column being 5678 into Counterpart which will become 0123.

The master Table 1 will show the group with the California Pick 4 number 0960. We will bring down the Master Table along with Master Table 1. If you are not sure how we recreated winning California Pick 4 number 0960, study Master Table 1 and pay close attention to the numbers in the 1st row as well as the group in the 1st and last column. Also note how the groups are placed to recreate 0960. You will afterwards graduate to Master Table 1.

Master Table

1	7	9	5
0	6	9	6
9	5	8	7
8	4	7	8

Master Table 1

0	9	0	6
1	9	4	5
2	8	2	4
3	7	6	3

We have so far recreated the California winning Pick 4 number 9587 in Master Table and also the 0960 as 0906 in Master Table 1. What is left at this point is the winning number 3261 to accomplish the task.

A closer look will show you that Master Table 1 already has 0123 in the 1st column position. The only missing ingredient is the number 6 that will make it 3261.

The tough part of the job is already done; creating the winning CA Pick 4 numbers 9587 and 0960 from a simple trend which presented most of the numbers we need for the 3261 in Master Table 2 (in the first digit column).

That task is so simple that all you need to do is switch the 0906 in the 1st row of Master Table 1 in the other direction. That switch will now turn the Pick 4 numbers in the 1st digit position from 0123 to 6123 thereby getting the highly sought after number 6.

We will bring down Master Table 1, switch the 0906 to 6090 to create Master Table 2.

Master Table 1

0	9	0	6
1	9	4	5
2	8	2	4
3	7	6	3

Master Table 2

6	0	9	0
1	9	4	5
2	8	2	4
3	7	6	3

We have now accomplished the goal of recreating the California Pick 4 numbers 9587, 0960 and 3261. You can do this exercise from different dates as well as lottery games in pursuit of the game you are betting on.

The more important thing is to convert Master Table 2 into winning Pick 3 numbers from another state as promised. That would be another surefire way to prove that the trend is intact and can be used to win both Pick 3 and Pick 4.

We are going to move the numbers from the 1st column of Master Table 2 and add them into the last digits to create the Pick 3 group and test them afterwards.

Master Table 2

6	0	9	0
1	9	4	5
2	8	2	4
3	7	6	3

The new minted winning Pick 3 numbers will be

090 + 6 =096

945 + 1 =946

824 + 2 =826

763 + 3=766

Master Table 3

0	9	6
9	4	6
8	2	6
7	6	6

We have now created winning Pick 3 numbers from Master Table 2.

Now let's put the group to test through recent Maryland winning Pick 3 numbers to see if they produced winnings which will be one more confirmation that the trend is in place.

Maryland Pick 3 lottery played 964 on 4/21/20, 826 on 4/22/20,906 on 4/24/20 and the trend completed with 766 on 4/28/20.

We have so far demonstrated that a simple trend can produce consistent winning numbers in Pick 3 and Pick 4.

The winning doesn't end there because all numbers are related and the key to betting outside the realm is by having the knowledge of N, S & C.

We cannot emphasize enough that all numbers are related and because of this singular reason there is no shortage of winnings once you apply the knowledge of N, S & C to complement that. The future winning numbers are hidden in the current winning numbers. We are sure that will sound far fetch. You can even pick any Pick 3 number off the top of your head and still zero in on winning numbers.

Let's demonstrate this in full view through the master workout section.

Master Workout

0	1	2
3	4	5
6	7	8
9	0	1
2	3	4
5	6	7

We will begin this exercise halfway in the middle of August 2019 with the winning number 974 and 390 that played on 8/1/19.

Master Workout			Master Workout 1		
0	1	2	0	9	8
3	4	5	3	6	7
6	7	8	6	8	2
9	0	1	9	7	4
2	3	4	2	2	1
5	6	7	5	4	5

We recreated the Maryland winning Pick 4 number by converting the middle digits into counterpart and then Shadow. It means that the 0 from 901 will be converted into Counterpart 5 (951) and then the 5 will be converted into Shadow 7, thereby creating 974. You will do the same conversion to the entire group that will produced Master Workout 1.

We will then proceed to recreate the next winning Pick 3 number 390 that played on the same day 8/1/19 with 974. This can easily be achieved by converting the numbers in the 3rd digit column of Master Workout 1 to produce 390 in Master Workout 2:

Master Workout 2

3	9	0
2	6	3
7	8	6
9	7	9
6	2	2
0	4	5

The trend is now set in motion because they continued by playing the next three wining Pick 3 numbers 656, 988 and 731 on 8/2/19 and 8/3/19. If you convert the Pick 3 winning numbers from Master Workout 2 into Shadow the entire group will now become the winning numbers in Master Workout 3. We will place the two workouts side by side for easy studying. Winnings like this should be enough reason to master N, S & C.

Master Workout 2			Master Workout 3		
3	9	0	0	6	3
2	6	3	8	9	0
7	8	6	5	2	9
9	7	9	6	5	6
6	2	2	9	8	8
0	4	5	3	1	7

The goal here is to show you different ways of hitting winning numbers. The same number you are looking at can produce several winning numbers only if you know how. We are going to place the same set of winning Pick 3 numbers from Master Workout in a different order, use it to recreate the same winning Pick 3 number 974 and other winnings you otherwise won't be able to see.

Master Workout

0	1	2
3	4	5
6	7	8
9	0	1
2	3	4
5	6	7

The above winning numbers will now be changed into the new set below:

New Set

0	1	2
5	4	3
6	7	8
1	0	9
2	3	4
7	6	5

With a little touch we now have the new set of winning Pick 3 numbers. The next step of course is to recreate the same winning Pick 3 number 974 that played in Maryland Pick 3 numbers of 8/1/19. The same Pick 3 number was created earlier in the master workout table. Let's do similar exercise to see what the outcome will be. In order to recreate the 974, the most visible entry will be from 678 because the 6 can be changed into Shadow making it 9 thereby having the numbers 9 & 7 towards creating 974. We will proceed to do that and worry about the 3rd missing number 4 that will make it 974 afterwards. Always remember that the numbers in the group exhibit similar behavior because they are in the same trends.

The group after converting the numbers in the 1st column will now change the new set into:

New Set 1

3	1	2
7	4	3
9	7	8
4	0	9
8	3	4
5	6	5

One possible candidate that will change the 8 of 978 into 4 making it 974 is to look at the numbers in the 3rd digit column. Those numbers are 2, 3, 8, 9, 4 and 5. The lottery house will love for you to choose the easy route which most people do and end up losing. We are not just going to place the numbers the way they are. We choose to change the group into Counterpart in order to shift the entire group 5 steps while keeping the highly sought-after number 4 in the midst to create 974. We also chose to place them after converting them into Counterpart in the other direction. The group 2, 3, 8, 9, 4 and 5 when converted into Counterpart will now become,

7, 8, 3, 4, 9 and 0. These new converted numbers will now replace the numbers in the 3rd digit column there creating the new trend with winning number 974 in New Set 2.

New Set 2

3	1	0
7	4	9
9	7	4
4	0	3
8	3	8
5	6	7

We have now accomplished the task of recreating winning Pick 3 number 974. With just the minor tweak the New Set 3 played all six winning Pick 3 numbers from the group in the same month. Take a look at the dates the entire group played.

974 played on 8/1/19

838 played on 8/14/19

657 played on 8/15/19

749 played on 8/16/19

304 played on 8/16/19

310 played on 8/22/19

You cannot ask for anything more. If you can put serious study on Lottery Maestro book, you can produce similar results consistently.

As we go along, we must accommodate those patient bettors who don't have time to work on different numbers or may prefer to bet on the same set of numbers until it drops. The best method for them will be the classic play that most often hit within a period of 2 months or less. Winning lottery number within a span of 2 months will be good for a lot of people. Some people actually bet on the same sets of numbers for about one year before they win. It could even be longer.

The Pick 3 lottery in most markets pay about $250 for a straight 50 cents win. If you bet straight and box on the same number for a period of one month, it will cost you about $30. If you win 50c box over that same period, you will get paid $40 and $80 if it's double number. If you win it straight and box, you will get paid between $290 and $330. In either case that will be excellent return on the $30 spent in that one month period. That alone will defeat the common axiom of 1 to 10000 winning odds.

The classic play actually exploit the weakness of the lottery numbers to give you room that could win a lot more over that period. They said that the patient dog get the best bone. In classic betting that would not be enough reward.

We actually believe that the patient dog do indeed get not just the best bone but the best meat with bone. After all the meat has some vital nutrients that leads us to classic play for the patient better.

Classic Bets

The classic play is where you divide the Pick 3 number for instance by 2 and add a different set of Pick 3 number to produce the likely winning Pick 3 numbers in 2 month or less. The classic formula is best explained by working it with actual Pick 3 winning numbers. We will work it out with 2 different states to show you the efficacy of the classic method. It is so good that many might consider adopting it. You can use it for as many as your pocket can carry you because the winning is consistent regardless of where you are betting.

Our example will begin with Maryland Pick 3 winning number from 5/1/19 results. It is imperative that we chose that date because the Pick 3 numbers in the classic bets method must be divisible by 2 but the actual result is 411. The 411 is definitely not divisible by 2. One option in this scenario will be to convert the 411 into Counterpart and it will become 966. Always remember that you get Counterpart by adding 5 to the number. The winning Maryland Pick 411 plus 555 equals 966. The converted Pick 3 number 966 is divisible by 2. The formula is,

966 divide by 2 plus 112 = 483 + 112 = 595

966/2 =483 + 112 = 595

The numbers you will be betting on are 595, 596 and 597.

The key number is 112. You add the key number 112 in vertical and horizontal order.

It means the 112 must be added to the result in the order below:

1	1	2
		1
		1

Alternatively, you can also look at it as:

1	1	2
0	0	1
0	0	1

The number 2 is constant which was why it created the 595 when added to 483. You only add the 2 once to which resulted in the new winning Pick 3 number 595 after which you will add 1 and 1 to create 596 and 596.

The final result from the classic method will be:

595, 596 and 597.

This method is for those that like to play the same set of number with the goal of winning within 2 months.

Now let's check the numbers against Maryland Pick 3 results to see what happened. Maryland lottery played 955 on 5/9/19. That would have been profitable since 955 is double number which pays $80 for every 50c box win. You would have won more than 3 times your cost. There are many that Hit straight which you will see later thereby paying you a lot of money.

You can leave the group and pursue another winning Pick 3 numbers using the same classic bets system. That takes us to the Maryland winning Pick 3 number 553 that played on that same day of 5/1/19.

The 553 is of course not divisible by 2, so we have to convert it into Counterpart making it 008.

Applying the same classic bets method:

008/2 = 004

004 + 112

116 + 1 = 117 + 1 = 118

The numbers you will be betting on are:

116, 117 and 118

Now let's take a look at the results.

Maryland played 811 on 5/25/19 which would have paid you more that your cost. So far the two Pick 3 numbers dropped in less than one month.

The next two winning Pick 3 numbers that played on 5/2/19 are 708 and 820. These two are divisible by 2 thereby showing you the lens from both angles.

Let's work on the next Pick 3 numbers 708 and 820:

708/2 = 354

354 + 112 = 466

466 + 1 = 467

467 + 1 = 468

The Pick 3 numbers to bet on are 466, 467 and 468.

Now let's take a look and see what happened.

Maryland lottery played 764 on 6/10/19 and that would be less than two months. It followed the tradition except that it played as box which won't pay a big payout. In any event if you worked on and played the next winning Pick 3 number from the same date, those numbers would be 522, 523 and 524.

820/2 = 410 + 112 = 522 + 1 = 523 + 1 = 524

If you had worked on and played just from the winning Pick 3 numbers above, you would have won straight on 524 that dropped on 6/13/19. Once again the 524 played in less than 2 months. We can now give hope to those that like to play the same set of numbers.

Patience should at this point be enjoying nice T bone.

The next Maryland winning Pick 3 number 431 played on 5/3/19

Follow the classic bet's method and see if you can work out winning number 506 that played on 5/10/19. Now that you mastered how to work out the winning numbers using the classic bets method, let's do the last Maryland winning Pick 3 number 498 from the same date of 5/3/19.

498/2 = 249

249 + 112 = 361 + 1 = 362 = 1 = 363

The numbers you will be betting on are,

361, 362 and 363.

Now let's check to see if any of the Pick 3 numbers above played over a two month period.

Wow! Maryland played 363 on 6/6/19 followed by 623 on 6/12/19 and the third one 361 played on 6/18/19. The three winning Pick 3 numbers played precisely 6 days apart, confirming how solid the classic method is.

Two of the Pick 3 groups played straight and all three Hit in less than 2 months. My friend at this point they need to throw in Porterhouse steak with the T bone and a bottle of champagne.

Let's take quick look at the Classic bet's power using Texas lottery. At this point there will not be too much explanation. If you don't understand it go back and study the Maryland lottery examples one more time.

Texas lottery played 778 on 7/1/19.

The Classic Bet's Method

778/2 =389 + 112 =501 + 1 =502 + 1 =503

The three winning Texas Pick 3 numbers from Classic Bets to play within the next two months are 501, 502 and 503

Texas Pick 3 played

520 on 7/2/19

305 on 7/8/19

503 on 7/9/19

502 on 7/11/19

520 on 7/13/19

052 on 7/27/19

520 on 7/2/19

035 on 8/9/19

501 on 8/29/19

You can see that the Classic Bets method produced so many Texas winning Pick 3 numbers over the two month period. Forget steak dinner, this one deserves owning a private jet.

The next Texas winning Pick 3 number on 7/1/19 is 111. In order for 111 to become divisible by 2, you have to apply Counterpart thereby making it 666.

666/2 =333 + 112 = 445 +1 = 446 + 1 =447

The Pick 3 numbers to bet on are 445, 446 and 447

The winning Texas Pick 3 number 111 followed its predecessor by dropping 644 on 7/10/19, 474 on 7/11/19 which would also be the same day that 502 played. The trend continued by playing 445 straight on 7/15/19 followed by 447 straight on 7/19/19 and also played 544 on 8/20/20.

At this juncture you have enough to prove that the Classic Bets method works. The betters who like to play the same set of numbers will have a field day.

The method works on every lottery number which means you can bet on more than one if you choose to or can afford it. If you are on tight budget, work on the numbers and wait for those that don't drop within the first 40 days and go for it from that point.

Some people pray to have just one winning number that will play within reasonable length of time. All you need to do is bring your creative mind and go for it.

The ball or sorry the winning number is in your court.

You have enough system to win all the Pick 3 games you want to bet on. The next chapter will bring another picture and also touch on the Pick 4 lottery.

Chapter 4

Lottery Winning Continuum

Winning the Pick 4 lottery, or any other game for that matter, cannot be done without making use of properly constructed tables along with the other methods we have discussed so far.

Let's construct a simple table and see how it stacks up with winning Pick 4 lottery numbers.

Table Construction

A	B	C	D	E	F
			3		
		2			
	1				
0					

Please note how we begin the table construction by filling the grid with numbers starting from zero. We are going to use 0 through 9 with a little tweak. The areas that you may find difficult to continue filling with numbers will show letters from A through Z as necessary. Afterwards they will be replaced with numbers until we fill in the entire grid and complete the table.

This table is highly important to winning every lotto game you can think of.

After placing the numbers 0 through 3 in the table construction above, naturally you would expect the pattern to continue with number 4--but we chose to continue from number 5 instead. That is done intentionally to help capture winning Pick 4 numbers that otherwise would not be there.

The process now continues with Table Construction 1

Table Construction 1

A	B	C	D	E	F
f		c		5	
	b		3		6
a		2		7	
	1		8		d
0		9		e	

The letters a through f (a=0, b=1, c=2, d=3, e=4, f=5) will be replaced with numbers in Table Construction 2.

Table Construction 2

A	B	C	D	E	F
5		2		5	
	1		3		6
0		2		7	
	1		8		3
0		9		4	

We have now completed half of the table. We will proceed to complete the rest of the table beginning with the number 6 since the last number we placed in the grid was number 5. Also note that the initial number 4 we omitted could be any of the other numbers. It doesn't really matter because you have the power of N, S & C to help you keep the trend in place at all times. The numbers that will fill in the rest of the table are as follows.

Table Construction 3

A	B	C	D	E	F
5	b7	2	f1	5	k6
a6	1	e0	3	j5	6
0	d9	2	i4	7	n9
c8	1	h3	8	m8	3
0	g2	9	l7	4	o0

The letters are only there to help you figure out how to construct the tables. Once the letters are removed you will have a complete table to work with. The complete table will be used to test Pick 4 winning numbers, which many find elusive. Once you see just how effective the complete table is, you will appreciate all the work that went into constructing it.

Those readers who wish to master how to construct more complex tables to help capture a lot more winnings should consider studying the ultimate lottery book, Lottery Icon.

Complete Table

A	B	C	D	E	F
5	7	2	1	5	6
6	1	0	3	5	6
0	9	2	4	7	9
8	1	3	8	8	3
0	2	9	7	4	0

Let's check to see if the complete table works using the Texas Pick 4 lottery.

Texas played 2792 on 9/2/19. A quick look at the complete table will show that you can easily recreate the Texas winning Pick 4 number from columns B, C, D and E where you will find the winning Pick 4 number 7215. The goal here is to apply N, S & C to the 7215 and recreate the winning TX Pick 4 number 2792.

Any conversion done to the 7215 must also apply to the group below the 7215 because they follow the same trend. The reason for that is to expose the other winning Pick 4 numbers that you may otherwise not be able to see.

The first digit number of the 7215 (7) can easily be converted to 2 by applying Counterpart. Remember that in Counterpart you simply add 5, reducing the number from 7 to 2.

The numbers in the 1st digit column are 7, 1, 9, 1, and 2 which will be converted into their Counterparts. This now changes the group to 2, 6, 4, 6 and 7.

The Pick 4 group has changed from:

B	C	D	E
7	2	1	5
1	0	3	5
9	2	4	7
1	3	8	8
2	9	7	4

To the new group:

B	C	D	E
2	2	1	5
6	0	3	5
4	2	4	7
6	3	8	8
7	9	7	4

The conversion so far is not looking too bad. The numbers under column C can also be converted into Counterpart to change the 2 to 7, resulting in a 27 that we can put towards creating 2792. You will also do the same conversion to the rest of the group under column C, after which you will now have:

The New Group 1

B	C	D	E
2	7	1	5
6	5	3	5
4	7	4	7
6	8	8	8
7	4	7	4

The next step will be to convert the number 1 under column D into 9, creating 279 out of our goal of creating 2972. The rest of the group under column D must also be converted to keep the trend intact.

The conversion will be changing the numbers into Shadow and then into Counterpart.

The numbers in column D are 1, 3, 4, 8 and 7. When converted into Shadow they become 4, 0, 1, 2, and 5.

The 4, 0, 1, 2 and 5 will then be converted into Counterpart which produces 9, 5, 6, 7, and 0.

The New Group 2

B	C	D	E
2	7	9	5
6	5	5	5
4	7	6	7
6	8	7	8
7	4	0	4

All we need now is to convert the numbers under column E that will produce the 2792. That requires once again converting the group into Shadow and then Counterpart.

5, 5, 7, 8, and 4 from column E when converted into Shadow will become 7, 7, 5, 2, and 1 after which we will apply Counterpart.

The 7, 7, 5, 2, and 1 after Counterpart application will become 2, 2, 0, 7 and 6.

The New Group 3

B	C	D	E
2	7	9	2
6	5	5	2
4	7	6	0
6	8	7	7
7	4	0	6

The winning Texas Pick 4 number 2792 that played on 9/2/19 is recreated with the accompanying group in the trend. Now let's check and see if some of the winning Pick 4 numbers played within a two month period. We prefer the methods to produce multiple winnings within a 2 month period so you can put some money in your pocket.

2792 played on 9/2/19

6552 played on 10/1/19

4760 played on 9/27/19

6877 played on 9/13/19

7406 played on 9/27/19

All the numbers in the group played within a one month period. I'd say that's very good for Pick 4 numbers that are supposed to play at a ratio of 1 in 10,000.

It is also important to recreate the same Pick 4 number in more than one area of the table simultaneously. It exposes the direction the game is playing as well as other winning numbers.

Complete Table

A	B	C	D	E	F
5	7	2	1	5	6
6	1	0	3	5	6
0	9	2	4	7	9
8	1	3	8	8	3
0	2	9	7	4	0

Complete Table 1

A	B	C	D	E	F
5	2	7	9	2	6
6	6	5	5	2	6
0	4	7	6	0	9
8	6	8	7	7	3
0	7	4	0	6	0

We created complete table 1 by inserting the new group that was just created under 2792.

You can create 2 additional tables that are completely different from the ones above and use all the tables to work on the same numbers. Use them for Pick 3, Pick 4 and other lotto games.

There are more ways to win the lotto games including the Pick 3 and Pick 4.

Chapter 5

The Number Switch System

The number switch system gives you a window to hidden lottery numbers. It is simple yet solid, and should be considered one of the best systems in the lottery world.

Take a look at simple Pick 3 numbers and what you can do with them by harnessing the power of the number switch system.

0	0	1
0	1	1
0	1	2

In the number switch system you always switch between numbers 1 and 2. Each one takes the place of the other and the trend shifts in that order towards catching the winning number you are working on. If both or only 1 or 2 appears in the group when you are using the number switch system you still need to switch the number.

Sample group			Numbers switched		
0	0	1	0	0	2
0	1	1	0	2	2
0	1	2	0	2	1

We are now going to form a simple group and switch numbers as necessary (with conversions) to catch some winning lottery numbers.

Before we delve into the Number Switch System, let us put down once more the N, S, & C conversions of 1 and 2. The conversions of the numbers 1 and 2 cover all numbers from 0 to 9 which is what you need to work on every lotto game.

See numbers 1 and 2 below with the conversions Shadow and Counterpart.

N	S	C	S	C	S	C
1	4	9	6	1		
2	8	3	0	5	7	2

As you can see the numbers 1 and 2 when rotated in N, S & C do indeed cover numbers 0 through 9. Pay particular attention to how 1 and 2 change into 6 and 7 when you apply Counterpart conversion. In Counterpart conversion you add 5 which means that 1 will become 6 and 2 will be 7. Those 4 numbers are central to the number switch system as you will learn later on in this book.

Now let's form a simple Pick 3 trend and use it in the Number Switch System to show you how you can capture more winning numbers. You can also use it in conjunction with the other methods mentioned in this book to win parlay betting. For those that don't know, parlay betting is when you are looking at getting 2 front or back pairs in your Pick 3 betting. The 50c win on parlay betting pays you about $25 in most markets. It primarily gives you the opportunity to play with the house money. A lot of players think that the $25 win is too small.

I once had 2 numbers that I played a $10 parlay on and the numbers hit. The payout was $500.

If you lay the methods outlined in this book side by side when you are working on your games, the winning numbers will appear more than once if you follow the instructions properly. In cases where you are sure of at least 2 numbers, you should consider playing the parlay. You can start with 50c and once your winnings become consistent, it's simply a matter of working your way up.

The Number Switch System

0	3	6
1	4	7
2	5	8

We will be using the above numbers to work on the Maryland lottery winning numbers that played from 5/1/19 which are 411 and 553. It is impossible to work out winning number 411 without applying the number switch method.

The Pick 3 numbers will become:

0	3	6
2	4	7
1	5	8

You always switch the numbers 1 and 2 before applying N, S & C conversions as necessary towards recreating the winning number 411.

0	3	6
2	4	7
1	5	8

#			S			#			C			#		
0	3	6	3	0	9	3	0	9	8	5	4	8	5	4
2	4	7	8	1	5	8	2	5	3	7	0	3	7	0
1	5	8	4	7	2	4	7	1	9	2	6	9	1	6

We now have Pick 3 number 916 that can be converted to create 411.

Once again the 841 represented the 553 in this trend. Instead of the group playing as 553, 771 and 185 it showed up as 841, 771 and 185.

Let's examine the role of the 841 by using the number switch system. Always remember to switch places between 1 and 2 until you get to a place where you can create the 841.

771	772	558	558	003	003	330	885	885	
553	553	770	770	225	115	447	992	991	
185	285	827	817	362	361	094	549	549	

The process continues by changing the 330 group into Counterpart despite the fact that the number 2 didn't show up. The same applies regardless. We are now at the point where the group can be converted to create 841. You can easily create the 841 from 549 but is designed for you to miss the trend. The road to the riches is well hidden.

A closer look will show you that the trend is hidden to the side because you can create 841 from 519 as well as 894. The 549 will produce a different trend which is why it's most readily apparent. The 519 and 894, on the other hand, could create the 841. Once tested, it will be in line with the trend to confirm that it represented the 553 we mentioned earlier. We are now going to switch the last group from 885, 991 and 549 into the order that will be easiest for most people to read and understand. They are the same group of Pick 3 numbers, but they can be switched between vertical and horizontal order as needed.

The winning numbers in the first digit column stay the same since we are creating 841 and the number 8 is there already. The 2nd column group (1, 9 and 9) will be changed into Counterpart to create a 4 that will go along with the 8 we already have. The last group (9, 4 and 5) will be changed into Shadow to create 1, to complete our target number 841.

Take a look at the group with 841 in it after the conversions. You can see the confirmation from Texas Pick 3 lottery that played 847 on 10/11/19. The fact that it took them several weeks to play the number still did not stop the number switch system from exposing it. The next winning number that played on the same date was 667. Some people might wonder why they played 667 instead of the 566 that we created.

I decided not to explain the reasoning on purpose, in order to give an extra learning opportunity to readers who want to study the lottery game further. Every number that plays has a reasoning behind it.

Now pay close attention.

As stated earlier, Maryland and Texas played the same group of numbers towards the end of July 2019. They both played 185 straight. Maryland on the other hand played 717 despite the fact that the system showed it as 771 which Texas played straight. The winning number 720 that Maryland played on 7/29/19 and repeated as 702 is part of that trend. If you take the middle digit which is 2 from 720 and place it in third position it will become that 702 we just mentioned. If you take the middle number 1 from 717 that played in Maryland and place it in the last position like we did with 702, the number will become the 771 that Texas played.

In order for the trend to complete, the middle number 4 from 841 must be placed in the last position to create 814.

The 566 in the trend below is the wrong number because the middle Pick 3 number 841 has the number 4 as its second digit. That is why Texas played 667 after 847 on 10/11/19.

You could pursue that which will follow a different trend but the problem remains unsolved. We cannot lose sight of our mission to prove that 841 was played as a decoy for 553.

The 841 switch in this case can be achieved by converting the 4 and 1 into Shadows which will produce 1 and 4, accomplishing our goal of creating 814. You also have to apply the same conversion to the other Pick 3 numbers in the group.

Please see below

0482	0481	3124	5124	5134
1593	2593	8760	7760	7740
2604	1604	4931	6931	6951
3715	3725	0587	8587	8577

We have just completed the process of recreating the Texas winning Pick 4 number 8577 that played on 10/1/19 which would have been impossible without employing the power of the number switch system.

The next step will be to recreate the TX winning Pick 4 number 6546.

The 6546 can easily be recreated from the 5134 group above. The winning Pick 4 number 6951 can be switched into 9516 for easy conversion.

When switched, the entire group will now become:

1	3	4	5
7	4	0	7
9	5	1	6
5	7	7	8

In order to recreate 6 towards creating 6546, the numbers in the 1st digit column will be converted into Shadow as well as the numbers in the 3rd column. After the conversion the above group will become:

4	3	1	5
5	4	3	7
6	5	4	6
7	7	5	8

You can see that the 6546 is recreated while keeping the 8577 in place. The proper order for the initial 8577 is actually 7758 that played on 10/25/19. It is not shown in the proper way which will help you win more, but it still couldn't escape the vice grip of the number switch system.

Texas lottery Pick 4 tested the actual winning Pick 4 number by playing 0806 on 10/5/19 followed by 7748. You can see that the 7748 is set exactly like the actual trend 7758. They came back exactly 20 days later and dropped 8600 followed by the winning Pick 4 number 7758. You will remember that the trend started by dropping 8577 which belongs to a different trend followed by 6546 and 2556 on 10/1/19.

In order to see where the trend is going, we will use the final workout and begin to recreate winning Pick 4 number 2556.

The winning Pick 4 numbers in the group are:

4	3	1	5
5	4	3	7
6	5	4	6
7	7	5	8

All you need to do at this point is use the number switch system on the above group and convert the entire group that will produce the given winning Pick 4 number 2556. The above group after applying the number switch system will become:

4	3	2	5
5	4	3	7
6	5	4	6
7	7	5	8

The new group above after conversion will change to:

2	5	5	6
6	2	7	6
3	6	6	4
8	3	3	0

The new 2556 group exposed the rest of the trend including the winning Pick 4 number 8330 that is among the group. Since the above group is worked out from the trend that had the actual winning Pick 4 number 7758 it shows that the 8330 is in fact a decoy for 8600. You should remember that it got tested earlier as 0806 with 7748 on 10/7/19.

The 8330 is the confirmation which is why the winning Pick 4 number 7758 played right after 8600 in Texas on 10/25/19.

Unearthing these winning Texas Pick 4 numbers was only made possible by making use of the Number Switch System.

Chapter 6

Winning Lotto System

Let us begin in earnest from the winning Powerball numbers that played on 1/2/19. Those winning Powerball numbers are 56, 46, 42, 12 and 08.

The first thing we are going to do is write down the individual winning numbers and convert them into Shadow and Counterpart to show their whole rotation. This is with the assumption that it is the only clue in your hand since they played as the first Powerball winning numbers for 2019.

We will examine the winning numbers individually to circumvent the astronomical odds of over 290 million to one. We can bring those odds down to one hundred to one by working out the individual winning numbers.

Let's go ahead and convert the winning Powerball numbers, 56, 46, 42, 12 and 08. The Powerball number is also part of the trend. The winning number 12 also played as the Powerball which is why we are not worrying about converting it twice.

Remember that we are filling the grids by converting the winning numbers into Shadow and then Counterpart through the rest of the rotation.

We will call the winning numbers in the first column the 56 group since it's the first number in the group and will use similar designations when necessary to describe any section you need to look at.

Lotto Number Rotation Table

56	79	24	81	36	09	54	17	62	98	43	10	65	97
46	19	64	91	46	19	64	91	46	19	64	91	46	19
42	18	63	90	45	17	62	98	43	10	65	97	42	18
12	48	93	60	15	47	92	68	13	40	95	67	12	48
08	32	87	25	70	53	08	32	87	25	70	53	08	32

The numbers in the first column above represent the very first winning Powerball numbers for 2019 and subsequent rotations where you convert them to Shadow and Counterpart.

Now let's begin the examination.

The next Powerball winning number 15 that played on 1/5/19 was among the next drawing. The numbers from the 56 group will give us the clue to what is going to drop next. You will find Powerball winning number 15 among the 36 group in the lotto number rotation table.

We have winning lotto number 46 that played in the first result. The number 46 and 15 appeared in the 36 column which confirms that the next Powerball result will play from that column. That number is 7/70 that played along with Powerball winning number 15. The goal is to capture three winning numbers based on prior winning numbers since they are in the same trend.

The winning number 7, however, shows in the rotation table as 70. It is important for you to note that the lotto numbers could play in either direction. Which means that the number 02 could play as 20 or the number 74 could also drop as 47.

In all honesty, you are not expected to make a big dent to the winning numbers at this point because you only had the 1st powerful result for the year to rely on so far.

The Powerball winning number 7 was actually playing from two different trends.

You will see the same number 7/70 on the 43 group.

The numbers from the 43 group are:

43, 64, 65, 95 and 7/70

The very first result for the year played the winning number 56 on 1/2/19 followed by the Powerball winning number 7 in the second result, which confirms the fact that the 43 group is playing.

They played 56/65, skipped the 95/59 and played 7/70. It goes without saying that the next result is going to play from the 43 group. They could go the opposite direction to play 43 or 64 but the most visible winning number is 95/59, which they dropped on 1/9/19.

This is what makes the lotto number rotation very powerful. The two winning Powerball numbers are given. All you need to do is follow the trend and play the next winning number. You don't have to fish for the winning Powerball number from 69 different numbers which will box you inside the pit of their high odds of winning. The prior two winning numbers already show you where the number is going to play from next.

The winning Powerball numbers from the 43 group are 56, 7/70 and 59/95.

Once you rotate enough numbers, you can easily pick the third winning number from the trend.

The next Powerball result on 1/12/19 repeated 7 and 36 from the 36 group just the same way they did with 7 and 15.

The number 7 in this case is playing for the 36 as well as the 43 group because the 59 served as a base once again for 7 and 56 from group 43.

See Powerball results for 1/12/19 and 1/16/19.

You can see how clearly the trend produced the required winning Powerball number for the next result.

After the number 15 that we worked on, the next winning trend will be number 27 that played on that same 1/2/19. Our job is to use it to figure out the trend and catch the third winning Powerball number.

There is no winning number 27 in the lotto number rotation table. If you converted enough numbers, you will see where winning number 27 will appear as well as the accompanying group like the ones we worked on. We are limited at this point to the five numbers that played on 1/2/19.

Those five will be enough to show you how to make good use of this lotto system.

What we are left with will be to see if we can recreate winning number 27 from what we have and see if it will be able to give us the third winning Powerball number that we are looking for.

That quest took us to group 81 because the winning number 25 can easily be converted to 27 by changing the second digits into a Shadow.

The numbers in the 81 group are 81, 91, 90, 60 and 25. Those winning numbers after the Shadow conversion will become 84, 94, 93, 63 and 27. You can see that we now have winning number 27 to work with.

The winning Powerball number 27 played on 1/5/19 followed by 49 on 1/9/19 and the third number we are looking for, 36, played on 1/12/19.

The two winning numbers 15 and 27 examined so far produced the third winning Powerball number we are looking for.

The third Powerball number that played on 1/2/19 is 69. We will continue to see if the number 69 will give us the third winning number in the trend.

The last conversion we did was from group 81 that had winning number 25 from which we created winning number 27.

The lotto number rotation table does not have the winning number 69. Since we did not convert enough numbers that will show winning number 69, our only choice will be to see if we can create the number 69 and hopefully that will give us the third number we are looking for.

That quest leads us first and foremost to group 98. The reason for choosing group 98 is because the number 25 appeared in that trend which means that it is closely related to group 81 that produced 27.

The Powerball winning number 69 can easily be created by converting the numbers in the 1st digit column to a Counterpart. The numbers in the 98 group are 98, 19, 10, 40 and 25 which after conversion will become 48, 69, 60, 90 and 75.

The next winning number in the group 6/60 played to complete the trend.

Powerball played 69 followed by 48 and 60 which we needed to win.

The three winning Powerball numbers played therefore defied the conventional odds.

The 4th winning number we are looking for is 7 which we discussed earlier. You can see the highly regarded number 7 played after 56 followed by winning Powerball number 59, which gives us the 3rd winning number we are looking for.

The last winning number to figure out is number 3.

The Powerball winning number 3 played from group 81. The winning numbers in the 81 group after conversion are 59, 69, 67, 97 and 03.

The winning Powerball number 3 and 69 played on 1/5/19 followed by 59 to complete the task. They played within two results because the number 3 is playing the exact same role as number 7.

The two numbers will eventually follow their own respective trends, but the method showed the numbers despite the fact that we had only one result for the year. The more results to look through, the easier the method actually becomes.

Your exercise at this point should be to convert the numbers in group 81 and arrive at the numbers above. The answer is already given. This exercise is for you to know if you understand the process. If you got the trend right, bravo, otherwise go back and study the material. This is not meant to be a quick or easy read. It is not a novel but the fact remains that once you master it, you will be better off.

The method shows that you are capable of picking the individual winning numbers without worrying about the high lotto odds.

Many readers may wonder why the group 81 was chosen to do the conversion to capture the winning number trend. You could convert some of the other groups and still get number 3.

How you approach this situation is to do rotation on the winning number 3 that will show you the group. Now let's convert the 03 into Shadow and Counterpart to show you why.

The conversion for number 03 from Shadow to Counterpart will be:

03, 58, 72, 27, 85, 30 and 03.

You can see from the rotation above that the winning number 3 is actually related to 27 and since we converted from group 81 to produce number 27, it is given that the 81 is playing the same trend that will expose winning number 3.

Do the above exercise before you progress.

Let's bring the lotto table down before we continue.

Lotto Number Rotation Table

56	79	24	81	36	09	54	17	62	98	43	10	65	97
46	19	64	91	46	19	64	91	46	19	64	91	46	19
42	18	63	90	45	17	62	98	43	10	65	97	42	18
12	48	93	60	15	47	92	68	13	40	95	67	12	48
08	32	87	25	70	53	08	32	87	25	70	53	08	32

The number rotations also produce winning numbers not just through the columns but rows as well. The rows format do often confirm more than one individual winning number. One such example will be to look at the one month anniversary of the Powerball result to see what happened.

The one month anniversary of the result for 2019 Powerball winning numbers for 2/2/19 showed:

10, 43, 17, 18, 65 and the Powerball number is 13

The lotto number rotation table above played a good percentage of those. Looking at the numbers from rows, go to row 42 and follow the numbers. You will see that row 42 produced winning numbers 18, 17, 43, 10 and 65. The same winning Powerball numbers were for 2/2/19. Those are all the winning Powerball results from just one row (42). That is unheard of in the world of predicting lotto winning numbers. It shows you that each of the rows is capable of producing all the winning lotto numbers with the comfort of figuring out the winning individual numbers through the columns.

You have to appreciate the opportunities you could put in your hand once you master how to use this method. The proper way to bet on lotto for most people would be not to play all the time. When you are ready to play, pick 18 numbers and wheel them, which will give you 42 combinations.

The numbers in the rows will be 12 and you still need 6 more numbers to complete the 18 before wheeling. This method doesn't require many numbers as you can see. It actually produces all the winning numbers in more ways than one.

The individual winning lotto numbers would be combed out from the columns as discussed earlier and the rows format will cement the deal.

If you want to study winning more on the rows, go and read the lotto section of the ultimate lottery book, Lottery Icon.

The lotto number rotation table is so powerful that it doesn't matter what lotto game you are betting on. As you may recall, the rotation is from the Powerball winning numbers of 1/2/19.

Let's use the same table to see if it works on the Mega Millions as well.

The Mega Millions seems to enjoy playing more of the numbers from the table.

Take a look at the Mega Millions result from 1/15/19 through the next five Mega Millions results that played all the five winning numbers in the column under the 54 group. We are not talking about getting the 3rd winning number here. It gave you the opportunity to have one winning number from the 54 group for the next five Mega Million results.

Lotto Number Rotation Table

56	79	24	81	36	09	54	17	62	98	43	10	65	97
46	19	64	91	46	19	64	91	46	19	64	91	46	19
42	18	63	90	45	17	62	98	43	10	65	97	42	18
12	48	93	60	15	47	92	68	13	40	95	67	12	48
08	32	87	25	70	53	08	32	87	25	70	53	08	32

There is a relationship between winning Mega Millions and other lotto numbers between different result dates. In order words, the first winning Mega Millions number in today's result has a relationship with the 1st winning number in the next result. The last result or any of the other winning results have direct relationships with the next result.

We are going to pick the 1st winning Mega millions number beginning in 2019 through the next three results in the same position, do some converting and use that to prove the point. We are essentially telling you that there are many ways of winning the lotto. This trend works regardless. You can go several months and the trend still works. The simple reason for that is because they are related.

Mega Millions played 57 on 1/1/19 followed by 29 on 1/4/19 and 13 on 1/8/19. Those are the very first winning numbers in each of the three Mega Millions results.

We used Texas lottery to check the past results because you can see results for an entire year under the same page and also you can check it in drawn instead of numeric order.

Lotto Number Rotation Table 1

57	75	20	83	38	02	57	75	20	83	38	02	57	
29	86	31	04	59	76	21	84	39	06	51	74	29	
13	40	95	67	12	48	93	60	15	47	92	68	13	

Make sure to look at the table carefully when you are working on catching the winning numbers. You can see that the first 75 group with winning number 86 must produce 40/4. You will notice that the second group 75 had number 84 instead of 86 which resulted in the third winning number being 60 instead of 40. You will also see that 83 followed by 4/40 must produce 67 in the trend while the same 83 followed by 06/60 instead of 04 must produce 47.

It is important because you might be looking at 83, for instance, and picking the wrong accompanying numbers without paying close attention.

There is no other way that the numbers rotate outside what you have in front of you.

Now let's see if the trend produced Mega Millions winning numbers.

The 1st 75 group with winning numbers 75, 86 and 40 could be seen on display when Mega Millions played 68 and 4/40 on 3/8/19. Those two winning numbers will alert where the numbers are playing from in the lotto number rotation table 1. The only winning number left to complete the trend will be 75/57 that played in the next Mega Millions result on 3/12/19. You can use the same process to expose all the winning numbers for the next result.

The process keeps on giving regardless of how long. It could go several months and the trend will still be in place.

The same trend continued when they played winning number 57 on 3/8/19, skip one result, dropped 68 on 3/19/19. All you need to do is skip one result and play the 3rd winning number 4 that hit on 3/26/19.

We jumped three months from January to March and produced consistent winning numbers based on trends from the 1st winning Mega Millions numbers in three different results. Now let's go another three months which will be June 2019 and see what happened.

Mega Millions played 68 on 6/4/19, 68 and 40 on 6/7/19, 57 on 6/11/19 and 40 with 57 played on 6/14/19 to complete the trend.

As you can see, they started by playing the Mega Millions winning number 68 at which point you won't know where they are playing from. However, they confirmed it by dropping the winning number 40 and repeated the number 68. The same trend continued by playing number 57 which they repeated in the next result with number 40 to complete the trend. Each of the winning numbers in the trend played twice in that same order. The winning lotto numbers in the same trend exhibit similar behavior.

This goes to show you that there is a direct relationship between the winning numbers in the same position over different result dates.

Lotto Number Rotation Table 1

57	75	20	83	38	02	57	75	20	83	38	02	57	
29	86	31	04	59	76	21	84	39	06	51	74	29	
13	40	95	67	12	48	93	60	15	47	92	68	13	

We have demonstrated that the lotto number rotation table can give you consistent winnings. We showed you how to use the tables provided to capture more winnings even when the winning number in question is not among the tables. You can do a conversion and still zero in on the winning lotto numbers.

One undeniable fact is that there are readers who will not put in any labor but want everything done for them. We cannot possibly offer you that level of comfort, but we will give you three lotto pulse tables that will help further.

In using the lotto pulse table you begin to look at the trend in which the winning number immediately shows up. You don't wait to have two winning numbers. A good amount of winning numbers among them will drop within two lotto results. The primary goal is to give you an idea of what is playing next. You can use them to shed more light on the lotto number rotation table you are working on.

Let us look at some examples with the lotto pulse table provided below.

Lotto Pulse Table

4	6	2	9	2	2	6	5	0	8
5	0	8	1	1	4	4	6	7	8
8	0	4	3	7	4	1	1	3	8

The Mega Millions lottery played winning number 50 on 5/28/19, followed by dropping winning number 8 on 5/31/19 to close the month. It also started the month of June 2019 by playing the Mega Millions number 46 to complete the trend. You can see all three winning numbers in the 46 group section of the lotto pulse table above. That would be the winning Mega Millions numbers that played in the first and second column of the lotto pulse table above.

You can check the same winning numbers from the beginning of that year.

We chose this section as the example for a reason. The three winning Mega Million numbers played right after each other. The lotto pulse table, however, is not constructed to necessarily give you winning numbers right after each other. That would be the role of the lotto number rotation table discussed earlier.

The lotto pulse table is primarily designed to give you a clue as to what is playing. The winning numbers from the lotto pulse do often play within two lotto results.

When you look a little further, you will see that the winning number 46/64 played on 7/9/19 followed by Mega Millions winning number 8 that played in the next two results. The remaining winning number in the group is 50 which didn't drop. If you followed the instructions by choosing the other numbers immediately after the number 46 played, you would have the number 8 as one of the winning numbers.

Lotto Pulse Table 1

4	8	1	6	3	9	6	8	2	9
4	8	1	8	6	4	5	0	3	7
3	9	1	4	4	0	6	0	2	6

In lotto pulse table 1, the 48 group is in the first and second columns. The winning numbers in the 48 group are 48, 48 and 39. It shows the winning number 48 as a repeat candidate. It does not mean that the winning number 48 will repeat all the time. It applies when the above trend is playing like the trend that started in 2/1/19.

The Mega Millions lottery played winning number 48 to start the month on 2/1/19 and repeated the number in two results later on 2/8/19 and completed the trend with winning Mega Millions number 39 on 2/12/19. This would have been a very good outcome since the lotto pulse will often give you two winning numbers in the group. In this case the entire group played. The system was able to capture the winning number 48 that repeated along with winning number 39 in the above lotto pulse table 1.

You must also pay attention to when a winning number is playing on another trend. The lotto pulse table has the winning number 39 in the 48 group that we just discussed as well as in the 39 group which has the numbers 39, 4/40 and 64. The winning number 39 played again on 3/1/19 and continued dropping the rest of the winning numbers in the 39 group. The winning numbers in the 39 group played as 39 on 3/1/19, 4 on 3/8/19 and 64 on 3/8/19. Those winning numbers played exactly two Mega Millions results apart from each other. The 39 trend and the rest of the group played through the whole time. That didn't stop the 48 group whenever the trend is playing like that of 6/4/19 when Mega Millions number 48 played, followed by 39 and of course the repeat number 48. The three winning numbers played in exactly two results apart from each other without losing sight of the repeat candidate which would be number 48. You have to recognize the pattern in play and follow accordingly.

The trend will always play regardless of how long. Take a good look through the rest of the months and you will even notice that the winning number 48 group played that trend from 10/1/19 when winning number 39 played, followed by number 48 on 10/3/19 and the repeat of 48 on 10/8/19.

If you put in a little work, you will notice where the winning numbers are playing from.

Lotto Pulse Table 2

5	7	2	3	1	2	1	5	0	9
4	3	0	6	8	0	0	2	6	3
6	2	0	7	2	5	4	8	5	9

The lotto number rotation tables are more precise to whatever lotto game you are betting on. The lotto pulse table on the other hand is there to give you the lending hand you may need down the line. You must also remember to use the Number, Shadow and Counterpart methods as necessary. The N, S & C makes it impossible for you not to see what is playing at any point in time. All it takes is a little effort to master it. If you are finding it a little difficult, concentrate on conversion of numbers 1 and 2 since they represent numbers zero to nine (0 through 9).

We expect you to have mastered it at this point. You will be losing more than 80% of the opportunities to capturing more winning numbers without it.

We attach all the importance to winning to it and as such we will put down the conversions for numbers 1 and 2 once more below.

The conversion, of course, is always in the order of Number, Shadow and Counterpart.

1, 4, 9, 6 and 1.

2, 8, 3, 0, 5, 7 and 2.

As you can see from the above conversion, the numbers 0 through 9 are represented which means that you cannot miss any number or trend once you master it.

You can elect to go through each chapter several times before moving on to the next. Some readers skim through it and don't quite grasp it. The information in this book is condensed. You will not understand it if you run through the pages like a novel because it is not. Check the numbers through actual state lottery winning results. You can also check the trends through other states that are not mentioned in this book. The winnings will be consistent because the trend is the key and when followed properly will yield good results.

There is no other lotto book that is capable of giving you the third winning lotto number based on what they played already.

There is no other lottery book that can work out winning trends based on one result or actually prove to you that all numbers are related.

The Number, Shadow and Counterpart system will always put the number right back in line.

You can put down a bunch of so called random numbers and work them back into winning lotto numbers by using N, S & C. That is how powerful the system is. You have not really started playing the lotto until you apply the Number, Shadow and Counterpart system.

You will learn how to construct the most perfect lotto table ever with the ultimate lotto key number. People talk about perfect this and that in the wheeling and other lotto systems. Most of them are repeating what others have said already with nothing to prove it. The key to every lotto will be exposed in the last chapter and above all you will get the lotto masterclass table with the maestro lotto key number.

Lotto Masterclass Table

5	7	4	1	6	1	0	6	7	4
3	6	6	2	8	8	7	5	3	9
1	4	4	2	9	3	1	5	0	7
0	5	2	3	2	0	9	8	8	9

The lotto masterclass table is as good as any lotto board or system you will ever come across. Our initial intention is to show you how to construct the lotto masterclass table coupled with telling you the lotto key number. The lotto key number is central to any lotto game you are betting on.

In the interest of keeping the Lottery Maestro book small and packed with lots of winning information we have decided to put down the Lotto Masterclass Table for you and also let you know at this point that the ultimate lotto key number is 13.

Teaching you the construction of the Lotto Masterclass Table with reasoning as to why the winning lotto number 13 is key will require an entire book by itself. We took this decision based on feedback from our numerous readers. The key thing we found out is that so many people don't want or like to read. This book requires studying and because of that we decided to limit it to half of the original size without compromising the potency of the information.

We also decided to include one more chapter that will teach you how to capture more winning numbers. The last chapter will show you a system that you will not find in any other lotto book. It has never been discussed or taught by anybody. We consider it so vital that we decided to just give you the Lotto Masterclass and use the rest of the space to discuss this new concept.

The Lotto Masterclass Table above will give you steady winning lotto numbers. It will consistently show you more opportunities to catch the winning numbers. If you combine it with the other methods discussed earlier, your limit will be untouchable.

Let us touch the winning lotto numbers from the first two digits of the Lotto Masterclass Table briefly. It is no different from the methods aforementioned, so we are not going to belabor the point.

The winning lotto numbers from the first two columns of the Lotto Masterclass Table fall under group 57. We will be looking at the trend through winning Powerball numbers before proceeding to the final chapter.

The Powerball winning number from any of the numbers in the 57 group could trigger a trend like the winning Powerball number 57.

The winning Powerball number 57 played with the number 36 on 1/12/19. The trend doesn't often play in this order. Your goal in each case is to know where the numbers are playing from and follow the trend as it plays.

You can clearly see the two winning Powerball numbers 57 and 36 that played among the Lotto Masterclass Table from group 57.

The next winning number that played was 14 on 1/16/19 and they completed the trend by dropping Powerball winning number 5 on 1/19/19. The 57 group gave you ample clues to capture winning numbers because Powerball lottery played all the winning numbers right after each other.

The group may not hit like that all the time, such as in the case of winning Powerball numbers 5 and 63/36 that played on 2/6/19 followed by winning number 14 that dropped two results later. The lead number 57 did not play this time around but you could have gotten the other winning numbers by making use of the other trends. You have several trends that will lead you to the same goal.

This, however, does not mean that the potency of the 57 group diminished or any of the other groups for that matter. You have to always pay close attention to recognize when the trend is playing. You can see good amount of winnings from the same group as well as winning numbers from the other groups.

Let us mention one place where the same 57 group played that a lot of players may not even recognize before moving on to the last chapter. We definitely would prefer Lottery Maestro to be no more than 125 pages because of the feedback from our readers.

A case in point would be the 57 group that played the trend from 8/24/19 through the rest of the month. Do count the winning numbers from the 57 group alone to get an idea of how many winnings you would have caught.

This could only be possible with efficient use of the Lotto Masterclass.

Chapter 7

Lotto Number Random Concept

The axiom about lotto and random numbers must be questioned. If you believe in lotto numbers being completely random like so many others do, you cannot come close to winning. Those that believe in random numbers will have to wait a long time before they can expect to see any winnings.

We do not believe in random lotto numbers, and we have indisputable proof to back our position.

This chapter will take the so-called random numbers and rework them into actual winning numbers. Once you master how to use it, you will have more than your share of winnings. The truth about lotto numbers are hidden and you can only expose them when you employ the powers of Number, Shadow and Counterpart along with all the systems described in Lottery Maestro.

This information is too vital to ignore.

Random Number Concept

You can pick any group of "random" numbers and use them to work out the winning trend in any lottery game. We are going to begin this exercise by using the following Pick 3 numbers,

0	3	8
2	5	6
7	3	0
1	1	4

The above numbers were chosen randomly. For this exercise, you can use any numbers off the top of your head or gathered by some other method. We are going to convert the above group of numbers to recreate the first winning Texas Pick 3 numbers from

June 15, 2019. You could choose any date and it wouldn't matter, which makes this exercise good for every lottery game.

You will convert the above group and recreate the first four winning Texas Pick 3 numbers.

Texas lottery played 701, 640, 589 and 105 on 6/15/19. The winning number 256 from the random group above can easily be converted to produce 701. You are not going to be that lucky in every case. In instances where you need to do more conversion, please follow the instructions outlined in Lottery Maestro.

You will convert the entire group because you are trying to turn the "random" numbers above into a winning trend. After converting the entire group to create winning number 701, the above Pick 3 numbers will now become:

5	8	3
7	0	1
2	8	5
6	6	9

You now have winning number 701 with the accompanying group. The next step will be to recreate the next winning Texas Pick 3 number 640. You cannot recreate 640 from the above numbers. It is important to figure out what to do in this situation. The above group needs to change into a form from which you can create winning Pick 3 number 640.

You need to have two numbers that could be converted into 6 and 4 towards recreating 640. The above group simply won't do it. The lotto number switch will not do it. The next option will be to employ the 1000 plan. The 1000 plan is where you subtract each of the above winning numbers from 1000. If you subtract the numbers and still don't have a way to create 640, apply Shadow and Counterpart conversion as necessary and subtract them again from 1000 until you expose the numbers that could form the 640 in question.

The previous Pick 3 numbers after applying the 1000 plan will become:

4	1	7
2	9	9
7	1	5
3	3	1

640 can easily be formed using some of the numbers in the above group. It can be formed from the 417 as well as the 299. In order to create the number 6, we convert the numbers in the third digit position with Shadow, which changes the group to:

5	1	4
6	9	2
7	1	7
4	3	3

The next step is to create the number 4, which means that all the numbers in the second digit column will be converted into Counterpart and the group now becomes:

5	6	4
6	4	2
7	6	7
4	8	3

We now have 642 in the group which means that the final step will be to convert the numbers in the third digit column. In order to turn the number 2 of 642 into zero, it will require changing 2 into Counterpart 7, then to Shadow 5 and lastly into Counterpart 0. You must do the same for the rest of the numbers in the 3rd column after which the above numbers become:

5	6	1
6	4	0
7	6	3
4	8	7

We have completed the process of recreating the Texas winning Pick 3 number 640. The rest of the group with 640 played primarily in any order over a two month period. It wouldn't have paid you enough money which is why this process requires recreating through the first four winning numbers. The more work you do, the more winnings you can create from these seemingly random numbers. What makes this formula formidable is that the lottery house doesn't know where you are working your numbers from and cannot avoid you as a result.

The next Texas winning Pick 3 number is 589 that played on 6/15/19. The 589 can easily be recreated from 487. We begin by converting the numbers in the last group into Shadow, which will turn the 7 of 487 into 5 (towards creating 589). After that the numbers in the 1st digit position will be converted into Counterpart which changes the 4 of 487 into 9. Remember that you always do the same conversion for the entire group to maintain the trend. After doing so the previous group will become:

4	6	0
3	4	1
0	6	2
5	8	9

The final step will be to recreate the winning Texas Pick 3 number 105. We did that by converting the 062. You convert the middle digits into Counterpart to change the 6 of 062 into 1, which goes towards creating 105. The last step is to convert the 2 of 062 into its Counterpart, 7, and then into its Shadow, 5, after which the above group will become:

1	4	7
9	3	9
1	0	5
3	5	1

We have taken random numbers and converted them into winning trends. The goal is to capture winning numbers after the conversions are completed. Let's check the last group but bear in mind that the other ones also produced winnings.

The final group produced about 5 winnings over a two month period, all of which played as box and would have paid you about $240; a slight profit over your betting cost of $180. That bet is profitable overall but not all that exciting. We don't want to bet $180 over a two month period to make $240. It may defy logic, but we want more. If you didn't, you probably wouldn't be reading this right now.

Is there a way you could win more with the same random numbers?

The answer is unequivocally yes.

You have to realize that winning numbers are hidden, and you can only expose them when you learn how to make proper use of Numbers, Shadow and Counterpart. The above winning scenario is designed for you not to dig further.

Let's put just one twist on the above final group and see what happens.

The winning numbers in the previous group are:

1	4	7
9	3	9
1	0	5
3	5	1

We are going to use the above group and recreate the very first winning number 701 that we started with. You are not expected to think this far ahead, and therefore not supposed to see the gold that lies ahead.

We can easily recreate 701 by converting the 5 of 105 into its Shadow which will change it to 7. After the Shadow conversion of the numbers in the last column, the new group will become:

5	4	1
6	3	9
7	0	1
4	5	3

The new numbers are completely different from the parent group. Now let's check the new group to see if it produced winning numbers over the same period of 2 months.

The new group played 541 straight on 6/18/19. You would have spent $16 at this point and made $290. The group before this could not be compared to this one by any means. You are able to capture 541 straight because you went above and beyond. You chose not to kowtow.

The new group continued by playing the trend at least six more times at half the time it took the other one. You would spend about the same $180 and made $780 instead of $160. That is the benefit of working your numbers outside the box.

We must not lose sight of the fact that these winning Texas Pick 3 numbers originated from "random" numbers. The numbers that produced the winnings are 541, 639, 701 and 453.

Texas Pick 3 lottery played 514 on 7/2/9, 396 on 7/3/19, 453 on 7/4/19 and 601 on 7/5/19. Take another look at the trend from the group that started with 541 against the above winning numbers Texas played. The winnings should be enough to convince you to join us in questioning the concept of random numbers in lotto betting.

There are so many ways to come up with winning lottery numbers. As I stated previously, you can pick any set of numbers off the top of your head. We are going to work with the three Pick 3 numbers below.

7	1	3
2	8	0
4	6	5

Once again it doesn't matter where the numbers are from or in the order they're in.

We used the above Pick 3 numbers to recreate Maryland winning Pick 3 numbers that played on 7/15/19 and 7/16/19. Those winning Pick 3 numbers are 276, 247, 062 and 360. We believe that you should know how to convert and recreate the winning lottery numbers in front of you. To that effect we will give you the recreated winning numbers in the next table.

The Pick 3 numbers in the first column (713 group) are the numbers from which the entire conversion was made.

713	276	247	062	360
280	723	752	327	023
465	341	998	613	917

The above table represents the conversion of the Maryland winning Pick 3 numbers that played on 7/15/19 and 7/16/19 from the 713 group. The more important point here is that you can use the set of numbers you chose randomly, convert them through any state past winning numbers and take the final result and do the same thing in a different state. Each time you do that, you will be realigning the random numbers into the actual winning trend. We have demonstrated earlier what you could do to capture more winnings.

We will use the winning Pick 3 numbers and do the same process through Texas Pick 3 winning numbers after which we will introduce you to how to capture more winnings.

If you are willing to put forth the effort, you can place each of the winning methods described in Lottery Maestro side by side. There is no winning lottery number that would not show up in your exercise.

Now we did the same process with the last winning Pick 3 numbers under 360 group through Texas winning Pick 3 numbers of 8/1/19.

As I said earlier, we will give you the table with the winning numbers already converted. The table with the winning Texas Pick 3 numbers is not meant for you to glance through. You still have the responsibility of doing the same conversion. As a matter of fact, you should close this book, do the conversion yourself, and see if you are doing it right or need to study the material further.

We do suggest that you study and understand each chapter thoroughly before moving to the next.

After converting the winning Pick 3 numbers from the 360 group through Texas winning Pick 3 numbers of 8/1/19, the above table will now become the new one below.

360	433	977	180	185
023	300	822	053	058
917	959	486	604	609

The above table represents the converted Texas winning Pick 3 numbers on 8/1/19. Let us look at different ways you can use the table to enjoy more winnings. The next winning Texas Pick 3 number after the conversion was 632 that played on 8/2/19. If we convert along with the group, it will expose other winning numbers in its own trend just like we have been discussing throughout this book. We just want to figure out if there's other ways to get more winnings.

The last winning set of Pick 3 numbers are 185, 058 and 609.

We are not trying to convert and produce the next winning Pick 3 number 632, but we also do not want to overlook its potential either. In looking for other ways to reach our goal, you may notice that the sum of the digits in 632 is 11.

Is there any way we can produce Pick 3 numbers with the sum of 11 to see if we can capture other winnings?

You can convert the 185 in multiple different ways that will produce a sum of 11. For instance, the 185 can be converted into 173 which adds up to 11. There are other ways but the most important thing is when it shows you an opportunity to capture some winning lottery numbers.

In order to convert the 185 into 173, the middle number 8 of 185 must be converted into its Shadow 2 and then into its Counterpart 7. The last digit of 185 must be converted into its Counterpart 0, and the 0 will then be converted into its Shadow 3. You also do the same thing to the rest of the group, which will change the trend to the set of winning Pick 3 numbers below:

1	7	3
0	2	0
6	8	1

Texas lottery played 317 two days later on 8/5/19 and then played 091 on 8/6/19. Let us give you enough benefit of the doubt that you might not have caught those two winning numbers. That withstanding, you can still definitely recognize that they are playing the new trend that we just created.

They returned to the trend a few days later when the 816 hit on 8/10/19 followed by the same winning Pick 3 number 091. Some readers may wonder why the 091 is in the mix since it didn't show up in our 173 trend. If you take the middle winning number 020 and subtract 1, you get 019. The 091 clearly represents the 020, and they gave you a clue towards that realization when it played earlier with 317. The number 001 played prior to 317 which is an indication of what lies ahead. If you subtract the 1 (001) from 2 (020) the remaining number will be 1. If you subtract 1 from 20 the remainder will be

019 or 19. The winning Pick 3 numbers are shown to you in a different order to prevent you from noticing them, but they cannot escape the N, S & C system. The 317 will produce an entirely different set of numbers but you are able to see it when you use the Number, Shadow and Counterpart method to expose the trend. They showed you 317 even though they are playing based on 173.

Once you master these methods, no lottery game can escape your iron grip. And you can take that to the bank.

They came back a month later and played lead group 185, 058 and 609.

The 185 played on 9/4/19 followed by 805 on 9/5/19 along with 002. You can see that they played 058 as 805 and 020 as 002. They are repositioning the numbers to create a different trend. The next winning number that played was 069 on 9/6/19.

The trends played right after each other, even though it took one month. But you still created another opportunity by using the sum of the lead number 185 to create more winning numbers.

A lot of winning numbers are hidden right in front of you until you think to work outside the box.

The next winning Texas Pick 3 number is 794 that played on 8/2/19. Let us show you three different ways to expose 794.

It's important to work your numbers back into the last trend when you are not sure. The reason for that is because all the numbers are related. That method actually exposes more winning numbers, as we demonstrated in the case that produced $500 more winnings over the same period by going back one step. We now have the new group,

1	7	3
0	2	0
6	8	1

We got to the above group by working on the Pick 3 number that will give us the sum of 11 for 632 which was how we worked out the 173 group above. You now have to work the 173 group back to produce Pick 3 winning number 632. Remember the numbers with answers to your questions are often hidden.

Converting the numbers in the 1st column to their Counterparts will change the previous group to:

6	7	3
5	2	0
1	8	1

The last step will be to bring the last digit numbers into the middle since we have 6 and 3 out of 632, and convert the middle digits with the Counterpart method which will change the above group to:

6	3	2
5	0	7
1	1	3

We have completed the task of working the trend back to winning Pick 3 number 632. The 974 can easily be exposed from this point because it is still part of the trend.

If you turn the above group to the side once more it will become:

6	5	1
3	0	1
2	7	3

A simple conversion of the group with Shadow will change 651 to 974 thereby changing the previous group to:

9	7	4
0	3	4
8	5	0

You will notice that the 850 was part of the group we started with. If you feel at any point that you are getting stuck, work your trend back to the last step which will often show you the route to take. You now have the 974 and the accompanying numbers in the group that gives you more opportunities to win.

Another way to expose trends is by realizing that you are not necessarily looking at the real picture all the time. A lot of bettors will tell you that they missed the winning number by one. If they understand the power of Number, Shadow and Counterpart, they would be able to eliminate that gap.

The original Texas winning Pick 3 number we started the conversion with was 433 that played on 8/1/19. Now let's work on the trend by subtracting 1 from the 4, thereby changing 433 to 333.

We are now going to reproduce the winning Pick 3 number 333 from our trend:

1	7	3
0	2	0
6	8	1

You will convert the numbers in the 1st digit position into Shadow which changes the above group to:

4	7	3
3	2	0
9	8	1

The next step will be to convert the numbers in the middle digit to Shadow and then Counterpart thereby changing the above group to:

4	0	3
3	3	0
9	7	1

The final step will be to convert the numbers in the 3rd digit into Shadow, changing the above group to:

4	0	0
3	3	3
9	7	4

As you can see, we are able to reproduce the winning number 974 that played on 8/2/19 in Texas by just subtracting 1 from 4 of the 433 and working with 333 instead.

The third step to recreating the winning Pick 3 number 974 is by moving to a completely different space. The N, S, & C methods will always bring the numbers in line, which is why you can work on the numbers with confidence once you master them.

The next winning Texas Pick 3 number on 8/1/19 is 822.

This time around, let's add 1 to the lead number 8 and change it to 9. The task now is to convert the same trend and produce 922 instead of 822.

1	7	3
0	2	0
6	8	1

The number 1 of 173 will be converted into Counterpart 6 and then Shadow 9 which changes the above group to:

9	7	3
7	2	0
4	8	1

You can see that the winning Pick 3 number 974 is already exposed. You could use all three simultaneously to show you which trend is playing.

Now let's go further and recreate the 922 instead 822 to see what it will expose.

You will convert the middle digit numbers into Counterpart and the above group will become:

9	2	3
7	7	0
4	3	1

The final process is to convert the numbers in the 3rd digit column into Counterpart first and then into Shadow and the previous group will become:

9	2	2
7	7	7
4	3	9

As you can see, the 922 conversion produced not just the Pick 3 number 974, but also the lucrative triple number 777 that played one week later on 8/12/19.

Some people bet on the triple digits every day throughout the entire year. You don't have to do that. The triple winning Pick 3 numbers are attainable through diligent practice and mastery of the methods outlined in this book.

Now let us raise the bar and use a set of so-called random numbers to see what happens. I expect to see success here, because I know that you can expose winning numbers even from the least expected places once you employ the Number, Shadow and Counterpart systems.

I know this because I have used it in several occasions where I was looking for winning numbers. I remember pulling into a mall several years back and the lottery was about to close. I thought of what to do and decided to write down several numbers randomly after, which I converted the section that would reproduce the last winning number. The next number after that was the triple Pick 3 number which I played. The bet was for $4 straight and $1 for the parlay which is a front and back pair. The Pick 3 number hit and the payout was $2050.

I didn't think much after that until the day I went to visit somebody at Sinai Hospital. I parked my vehicle and looked at the Sinai Hospital sign.

It suddenly occurred to me to use an alphanumeric system. I converted the hospital name and applied Number, Shadow and Counterpart. It gave me winning number 4040, which I bet on.

Lo and behold, that number played. That type of winning Pick 4 number pays good money even if you only win the box. You can learn about the alphanumeric system through Lottery Icon.

These anecdotes are reason enough to question the concept of random numbers when it comes to lotto. All numbers are related and can easily be called back into the trend once you apply N, S & C.

I will now put down a set of random numbers in no particular order, and use the numbers to form a table, after which we will convert one section into actual winning state lottery Pick 4 numbers to see what happens.

The random set of numbers below are,

893027651033419

276349812204073

95357293147638

3490281

The above are numbers randomly called out. We will now use the numbers to form a lottery table. Study the completed table which shows you how to construct lottery number tables.

You can use the table for Pick 3 as well as Pick 4. Make use of it in conjunction with the other methods discussed.

The randomly called numbers I listed above have been transferred to the table below:

Random table

8	4	2	9	3	7	2	3	6
3	2	1	7	2	7	6	4	9
1	4	0	0	5	5	4	9	5
3	4	3	3	1	1	2	0	0
0	9	5	0	3	7	6	2	8
8	9	3	9	6	7	9	8	1

Now that we've formed the table from the randomly called numbers, it's time to test to see if it will produce winning numbers, or at the very least direct us to where the trend is playing. We have already shown you how we used random numbers and zeroed in on Texas lottery winning numbers.

We are going to use the table for a different state's lottery results. Let's use it on California Daily 4, or what most places call Pick 4.

We looked at the past California lottery results and can only go as far back as 11/23/19 that shows winning California Daily 4 winning number 7313. We will now proceed by looking at the above table and figure out where to recreate the California winning Pick 4 number 7313.

The table shows that it can easily be formed from numbers 2, 3, 6, and 8. You will find 2, 3, and 6 in the last three columns to the right and the number 8 will be found in the first column. Always remember that the entire table goes through the same transformation.

To make it easier, we will put the previous numbers down below before we continue.

2	3	6	8
6	4	9	3
4	9	5	1
2	0	0	3
6	2	8	0
9	8	1	8

We can easily form the California winning Pick 4 number 7313 from 2368 by converting the number 2 of 2368 into Counterpart 7 (7368) and continue by converting the 6 and 8 into Counterparts 1 and 3 respectively, changing the 2368 to 7313. We will do the same exercise to the rest of the group. Remember that the numbers converted here are the first, the third, and the fourth digit numbers in those columns through the rest of the group. This will change the original table of random numbers into:

7	3	1	3
1	4	4	8
9	9	0	6
7	0	5	8
1	2	3	5
4	8	6	3

California Daily 4 played 7313 on 11/23/19 followed by 3648 on 11/24/19. The converted winning numbers above produced winning number 4863. That would be the last winning Pick 4 number among the above group. The winning Pick 4 number 4863 is the same number California daily 4 played on 11/24/19. As you can see the random number produced the two winning numbers effortlessly. This is the same process that produced the two numbers that I won with last time.

This may not happen all the time but it will definitely set you in the right direction. There is also a good possibility that if you put down several random numbers and converted them, you could produce a whole lot of winnings.

When you work on Pick 4 with the different methods we discussed so far, you might come to a situation where you are sure of three out of the required four numbers from one system and get the missing number from the other one. You observe the winning converted numbers carefully against what is playing to produce the next potential winners. One such situation will be in the numbers 1235 and 7058 above. California lottery played 7325 on 11/30/19. That winning Pick 4 number has three numbers in common with the 1235 that the random number produced. You will also notice that California played 9850 on 12/2/19 and our group has winning number 7058 which shares three common numbers.

If you convert the numbers in the first digit position for 1235 and 7068 against actual California winning numbers those two will become 9058 which California dropped and the 1235 will produce 8235 after conversion while California played 7235. In this scenario you clearly have the shared numbers, and simple conversion produces one of the winnings. The other winnings might be caught if you use the other methods in conjunction with this one.

You can also repeat the conversion like we discussed on Pick 3 (i.e. going back and repeating it) which could produce more winnings or expose the missing digit.

You will also notice that similar numbers played as 7699 and 8914 on 1/7/20 and 1/8/20. In the above random group we have winning numbers 1448 and 9906. If you convert the 0 of 9906 and convert one of the fours of 1448, you will produce 1498 which California played; the other one will produce 9956 which is close to the other number California played, 9976. Just take this as more evidence supporting the use of multiple methods at one time.

The trend against what actually played is one more reason to question the veracity of the random number concept.

The first two winning Pick 4 numbers from the previous group are:

7313 and 1448

Let us convert it into the California winning Daily 4 number 9850, just for demonstrative purposes. To make things easier, we first rearrange the above winning numbers to 1373 and 4814. The number 1 of 7313 will have to be converted to its Shadow and then it's Counterpart to produce 9, while the 3 in the last digit position will be converted to its Counterpart to produce 8. The remaining two numbers, 7 and 3, will be converted via Shadow to produce 50. After conversion the above two winning Pick 4 numbers will become 9850 and 6314. California lottery played 9850 and 5314. In this instance, one simple conversion produced three of the required four numbers.

The last two columns of the random table produced 9508 and 4902 while California Daily 4 played 9508 and 3902 on 12/1/20 and 12/2/20. Look at how close we already are without even doing any conversion. Once you master how to build the complete table, choose a different set of numbers and do the same thing. This is betting beyond the realm.

The consistency cannot be overlooked.

We are going to do one more conversion for the doubters who are reluctant to join us in questioning the concept of random numbers when it comes to lotto. We are going to use one part from the same section of the allegedly "random" table that we used to work out the California Pick 4 numbers. This exercise will involve converting from that same section to create trends from the California winning Pick 4 number played on 11/23/19 to winning Maryland Pick 3 numbers using the random table. We want to know if it will produce winning Pick 3 numbers, with that date of 11/23/19 being the benchmark.

Section of the "random" table

2	3	6
6	4	9
4	9	5
2	0	0
6	2	8
9	8	1

Maryland Pick 3 lottery played 114 on 11/23/19. We can recreate the Maryland winning Pick 3 number 114 from 649. The first step will be to convert the numbers in the first column of the table to their Counterparts, which will result in the table pictured below:

7	3	6
1	4	9
9	9	5
7	0	0
1	2	8
4	8	1

The next step will be to convert the numbers in the middle column using the Shadow method, which then creates the following table:

7	0	6
1	1	9
9	6	5
7	3	0
1	8	8
4	2	1

We now have a row with 1 and 1, which we can put towards recreating the Maryland winning Pick 3 number 114 that played on 11/23/19. Do not forget 11/23/19 was the same date that the supposedly random table captured two winning California Daily 4 numbers. Now that we have 1 and 1, the final step will be to convert the numbers in the last column into Counterparts so that the number 9 will become 4. After conversion of the numbers in the third column, the above group will now become the final one pictured below:

7	0	1
1	1	4
9	6	0
7	3	5
1	8	3
4	2	6

The final set of Pick 3 numbers we will be checking to see if some of them played are 701, 114, 960, 735, 183 and 426. These are the same Pick 3 numbers from the table we just created. We are trying to see if one simple conversion using the power of N, S & C will transform the group from something "random" into one which displays an actual winning trend with two different lotto games from two different states.

Maryland Pick 3 lottery played 357 on 11/5/19, followed by 906 on 11/7/19, and 735 which hit the next day on 11/8/19. You can clearly see how close those winning Pick 3 numbers played after each other. The Maryland winning Pick 3 number 735 that started the trend came back and played as 375 on 11/19/19, signaling to you that the other winning numbers among the group are repeat candidates. You wouldn't know about the numbers in the group without having the original "random" numbers.

The winning numbers play so close to each other because N, S & C is forcing the entire group into the winning trend. The Maryland lottery Pick 3 key number 114 played on 11/20/19 which is one day after the 375 that played on 11/19/19, which serves to further bolster our argument that all numbers are related. The winning numbers so far married into the same 114 clan, so by default they must follow the same order. The Maryland winning Pick 3 number 690 repeated on 11/22/19, followed by the same key Pick 3 number 114 as a repeat on 11/23/19. They also played just one day apart, and that in and of itself is another confirmation that N, S & C have put the entire group back in line so that you can enjoy more consistent winnings.

Two days later Maryland Pick 3 played 462 on 11/25/19 and repeated it as 426 the next day on 11/26/19. You shouldn't need a fortune teller at this point to tell you that the group does indeed repeat. The Maryland winning Pick 3 number 183 wasn't left out either; it played two days later on 11/28/19. It also played straight like its predecessors, and that pays out huge money. The key winning Pick 3 number 114 continued the trend when it played as 411 on 12/12/19, followed by 381 on 12/21/19 and the winning number 170 that played on the exact one month anniversary of the key number. That anniversary date would be 12/23/19, and winning number 375 followed shortly behind, dropping the next day on 12/24/19. The results really do speak for themselves.

You can see that all the Pick 3 numbers in the group played over a one month period, with some in the group dropping several times. These winnings all originated from the same section of that so-called "random" table. That same section produced the California winning Pick 4 numbers and the Pick 3 Maryland winning trend, which caught more than 13 winning numbers over a one month period with a good percentage of them playing straight.

You can equally recreate trends from the same table towards the middle of the month with the numbers that played earlier, confirming the trend.

One area that is consistently overlooked by many is betting on the parlay, which gives you an opportunity to win when two numbers show up together. It gives you more avenues to win when you are betting on Pick 3 and only two of the numbers you bet on show up. A lot of people often miss the third number. The parlay, however, will pay you some money if/when this happens, rather than you losing out completely. In most places you get paid $25 for every 50 cents win. If you look at it from the eye of a scholar, the odds drastically tilt in your favor when making a parlay bet. If you can bet less than $25 and win you will be turning a profit. Do not look at it from the layman's point of view where you think that $25 isn't enough, because if you can win $25 repeatedly it starts adding up. The belief from the other side of the fence is that you will bet $50 to win $25. Our methods will show you how you can bet less than $12 and win more than double that by turning the table against the house.

We touched on the subject of parlay betting briefly earlier but decided to go through it once more because it gives you the opportunity to play with the house's money.

We are going to do one demonstration by taking a section of the Complete Table from Chapter 4. We will be using winning numbers from the last three rows of the table in columns A through F.

See the winning numbers from that section of the complete table and how we will use them to win a parlay (or what some people call a front and back pair) below:

From Complete Table

0	8	0
2	1	9
9	3	2
7	8	4
4	8	7
0	3	9

California lottery Daily 3 played 119 on 12/4/19. The winning Pick 3 number 219 is among the numbers we have from the Complete Table section above. The 219 is extremely close to the 119 I just mentioned. This means that you should pay attention to see if that group is trending, which incidentally happens to be the case. It is trending because California Daily 3 played 249 that same day. The winning numbers at this point will be 119 and 249 while our group has the winning numbers 219 and 932. The trend clearly shows you that the next winning Daily 3 number is going to be something close to the 784 in our table.

Since the prior two winnings had only two numbers from our group, it goes to show that playing parlay will be a good idea based on our track record thus far..

You would have won in this case because California Daily 3 played 842 while our system has winning number 784. You would have won with the front pair (84X) and get paid $25 for every 50 cents won.

Your total cost on the bet would be $6 and that's less than 25% of your winnings. That definitely tilts the game in your favor, and you are just following the trend shown to you by the house.

If you look through a few days' results you will notice that California played 239 on 12/8/19 and we have that number in the complete table from earlier. The winning Pick 3 numbers 015 and 400 played before and after the winning number 239.

We will not bet parlay on it because you need the second winning number to confirm the trend. Those winnings could be exposed if you are looking through more than one trend at a time. It is always advisable you use more than one trend. Once you master that, you can win parlays all the time and eventually increase your bets to win more money.

There are other parlay opportunities that are meant to be difficult for you to notice. As you go along and understand the methodology better, those are the ones that you will be betting more money on. They never fell because it takes time to recognize it.

One of those types of opportunities was when California Daily 4 played winnings numbers 620 on 12/11/19, 418 on 12/12/19 and 980 on 12/13/19. You will not notice it from the complete table section we are using but those winning numbers are there. That parlay group is represented by the winning Pick 4 numbers 080, 039 and 487.

Before they played 080, they played 620 as a decoy first. The sum of the digits in 620 is 8, just like the sum of the digits in 080 is 8. That number initiated the trend. The next winning number in the group that played was 418 which you will find has two numbers in common with the Pick 3 number 487 from the Complete Table. In other words they played 080, skipped 039 and dropped 487 then finally came back and played from 039 in order to produce 980 and complete the trend.

Anytime you recognize a decoy in the trend, seriously consider betting on it. It rarely fails.

We are going to expose some hidden opportunities with the same group of winning numbers from the complete table section. After all, the goal here is to make you a more formidable bettor.

We started that group on getting a parlay from the beginning of December 2019 and will finish with the same group for the year 2019, then start off the year 2020 profitably.

From Complete Table

0	8	0
2	1	9
9	3	2
7	8	4
4	8	7
0	3	9

The California winning Daily 3 numbers that played at the end of 2019 and started the year 2020 are 036 and 365 on 12/30/19, 717 and 302 on 12/31/19, 372 and 651 on 1/1/20.

Once you convert the group, the window to winning the trend is to see two winning numbers in the group, one of which might play as a decoy to throw you off. We showed you one example of such an occurrence earlier.

In the above group if you recreate the winning number 063, you would not be able to get the pairs lined up to capture the third winning parlay. Instances like that are bound to happen, but if you pay attention to the explanation here, you will be able to overcome such a deficit.

We chose to convert the group in a way that one of the numbers will give us a sum of 9 because the winning Daily 3 number 063 we are working on has the same sum of 9.

Always remember that the easy route is meant to throw you off of any possible opportunities to win big. In the above group, we could easily create a Pick 3 number that will give us the sum of 9 by converting the numbers in the 2nd digit column, thereby creating 009 from 039.

We chose instead to recreate it from the hidden number 932 which is somewhat in the middle among the group. You will succeed more by developing the discipline to avoid the easy route. That process will begin by converting the numbers in the middle digit column into their Shadows, which will change the number 3 of 932 into 0. Always do the same thing to the rest of the group. The next step will be to convert the 2 of 932 into

Shadow 8, then into Counterpart 3 and finally into Shadow 0 to change the number 932 to 900, accomplishing our goal of creating a Pick 3 winning number that will give us the sum of 9. Once you put the rest of the group through the same process, you will now have the Pick 3 winning numbers in Step 1:

Step 1

0	2	2
2	4	4
9	0	0
7	2	9
4	2	3
0	0	4

The next step will be to recreate California winning number 365. Always remember that the concern in working the parlay is coming up with two common numbers and not necessarily 3.

We found out that recreating the winning Pick 3 number 365 will not give you what we are looking for. We want to end 2019 and start 2020 off with a bang, so we chose once more to convert the numbers in the above Step 1 trend that will produce a Pick 3 winning number with the sum of 14, just like the winning Pick 3 number 365 in question. That leads us to Step 2 on the next table:

Step 2

0	7	2
2	9	4
9	5	0
7	7	9
4	7	3
0	5	4

The winning Pick 3 number 950 has a sum of 14 just like the winning Pick 3 number 365. The conversion to create Step 2 is simply done by converting the numbers in the middle digits to their Counterparts. We have so far represented the winning numbers 063 and 365.

The final stage is Step 3, where we moved on from creating the sum to recreating the actual winning California Daily 3 numbers.

The next winning number in the trend is 717.

We could recreate 717 easily from the 779 of Step 2, but that will box us into the trap of taking the easy route. Instead, we choose to recreate the winning number 717 from 473. Yes you read that right. It's certainly more work, as it requires more conversions, but this is the kind of discipline and keen eye you need in order to produce more winnings. We switched the 473 to 743 and did the same to the rest of the group.

We believe by now that you understand how to use N, S and C to arrive at the numbers you are working on. The simple description of the process is that we converted the 4 of 743 into its Shadow 1, which made it 713. After this, we converted the number 3 into Shadow 0, then Counterpart 5 and finally into Shadow 7, finally creating the winning California Daily 3 number 717.

This gives us Step 3, from which we will begin to bet on the parlay.

	Step 3	
7	3	0
9	8	9
5	6	2
7	5	4
7	1	7
5	3	9

We can now enjoy several parlay winning numbers like the common pairs 302 with 730, 189 with 989, 365 with 562, 651 with 562, 372 with 730, and of course 717 that would pay you on straight, box and parlay bets. The 717 alone, if properly bet on, will cost you $4 to place the bet, and pay you $380 in most lottery markets. In the interest of time I won't go into specific detail about the other winnings, but rest assured that they would only add to that.

That's definitely a good way to end 2019 and start 2020.

What the above exercise is intended to teach is that there are more winning numbers to be discovered if you acquire the patience and discipline to put in the work. You could win a measly $25 parlay by choosing the easy route, while somebody else is harvesting hundreds more winnings from the same trend.

The parlay is so simple to partake in that even the random numbers can produce parlay winning numbers.

We cannot count the number of winnings over the same time period from the random number table below.

Use the following random table to figure out winning California Daily 3 numbers from 12/30/19 through 1/10/20. You will be amazed at how many winning numbers you can fish out before you even put in any work or do any conversions. After that, use the same table to zero in on the next three winning parlays from any state.

Random table

8	4	2	9	3	7	2	3	6
3	2	1	7	2	7	6	4	9
1	4	0	0	5	5	4	9	5
3	4	3	3	1	1	2	0	0
0	9	5	0	3	7	6	2	8
8	9	3	9	6	7	9	8	1

Go back through "Lottery Maestro" as many times as necessary. Some of the information in this book will be new to so many of you, and it may seem overwhelming, but diligent effort will keep you in the game. Do not let lazy people drag you down under the guise of being "experts". Some of these people do not give you actual reflections on the material because they don't fully understand it themselves. We give you actual dates and where the numbers are playing from, so you can check it out yourselves. We also encourage you to check how many winning numbers you get before and after studying Lottery Maestro. We hear from readers all the time, who put in the necessary efforts and ultimately reap the rewards. Some readers want all the winning numbers to be served to them on a silver platter. But the truth is, there is no quick fix; Lottery Maestro certainly isn't one. But one thing I can authoritatively tell you is that there is no lottery book on earth that's efficacy can come close to Lottery Maestro.

The systems in this book have been honed and tested obsessively over the years.

Lay out different methods side by side. Take your time when you are working on the numbers.

The winnings will come after the mastery of the systems discussed in Lottery Maestro.

The Number, Shadow and Counterpart system is your best ally when it comes to betting on the lottery. Once you master that, there is no lottery game that can escape your iron grip.

The more you study Lottery Maestro, the better you become. You will find out that you gain more knowledge each time you study the material.

As stated earlier, if you find the material a bit difficult, stay on that chapter until you master it. Feel free to go through that same chapter several times. There is always another day to win. The cost of rushing through the book like a novel is more than the gain of winning by diligently mastering the systems in Lottery Maestro.

You will be betting outside the typical realm once you study Lottery Maestro and bring in the discipline and patience to go with it. Also remember that checking the methods using the actual results is very important. There are no shortcuts. People who try to take shortcuts in life find themselves facing the consequences, and studying Lottery Maestro is no exception.

The potential payout in lottery games means that you have to do some work on your part. But at least the hard part of learning the techniques is already done for you.

Chapter Supplement

Lottery Maestro Signature

The common belief that you have to arrange numbers in numeric order does not hold water in the game of lottery. In fact, the opposite is true. If winning numbers are worked out in numeric order, nearly every one would win endlessly because that is how people are programmed to think.

The "Lottery Maestro Signature" is here to demystify that myth. The ultimate secret not known to bettors out there is that all numbers do, indeed, move in the quest to capture winning trends. I will demonstrate this with actual winning Maryland pick 3 numbers. The wonderful news is that it works on Pick 4 as well as every other lotto game that's played.

Once you have the basic trend to follow what is playing, you will realize that each winning number gives you the clue to what the trend is playing. You can actually work out all the winning numbers over that month with that basic trend. You can also use the same trend and catapult it to other winning trends in any state or country of your choice and still capture consistent winnings.

I will use numbers 0 through 9 to form a basic trend from which I will demonstrate capturing Maryland winning Pick 3 numbers in May 2019. Feel free to use the same basic trend and apply the methodology to any month or year you choose to see the efficacy right before your eyes. Once you master the "Lottery Maestro Signature", the winning lottery numbers will be in your court.

Basic Trend 1

0	1	2	3	4	5	6	7	8	9

Basic Trend 2

0A	1B	2C	3D	4E	5F	6G	7H	8I	9J

I will now form the basic trend construction with letters to show you how the numbers are placed before I show you the final trend. This exercise is meant to help you do it on your own.

Basic Trend 3

A	J	I
B	F	E
E	D	E
C	G	D
A	H	B

The above letters are meant to show you the order the numbers will be arranged to help capture lottery winning numbers.

Basic Trend 4

0A	9J	8I
1B	5F	4E
4E	3D	4C
2C	6G	3D
0A	7H	1B

The basic trend 4 are basically the same set of numbers from basic trend 2 above. The letters are meant to help you further understand how I come up with the winning trend. Please do this exercise on your own and also note that you can use the same format to place entirely different sets of numbers because the "Lottery Maestro Signature" will zero in on the winning numbers regardless of the order.

(01), (432), (5), (67), (98), (01), (23)

Signature Trend

0	9	8
1	5	4
4	3	2
2	6	3
0	7	1

The signature trend, in this case, is used to work out Maryland winning Pick 3 numbers for May 2019 beginning with number 411 that played on 5/1/19. Remember from the onset that "Lottery Maestro Signature" can use any and all the numbers in the trend as necessary to capture the winning trend. In other words, you can switch between any and all the numbers as necessary. If it becomes necessary to switch only the lead numbers or the middle and last digit numbers do so towards exposing the winning number.

The signature trend does not have any group that can produce the winning Pick 3 number 411. I will begin the process of recreating winning number 411 by changing the numbers in the middle digits from signature trend into Counterpart after which the above signature group will become:

0	4	8
1	0	4
4	8	2
2	1	3
0	2	1

I have now produced 104 among the group that is closer to 411 except that I will prefer the 104 to be in the form of 401 that is a lot closer to 411. I can easily change it to 401 by applying Shadow to the 1st and last digits thereby changing 104 to 401. Always remember that you will do the same conversion to the rest of the group because they belong to the same signature trend. After converting the entire group by applying

Shadow to the 1st and last digit numbers, the above group will now change to the new group below:

3	4	2
4	0	1
1	8	8
8	1	0
3	2	4

The above group is a lot closer to creating 411 that Maryland played on 5/1/19. Since "Lottery Maestro Signature" enjoys the luxury of using any numbers in the group, the best place to recreate the 411 will be with the middle Pick 3 number 188 in the group. The 1st digit number 1 of 188 can easily be converted to 4 by using Shadow. It is very tempting to try and recreate it through the winning number 401. That would precisely be what the lottery house expect you to do. The "Lottery Maestro Signature" takes you through the least expected route which is the reason why it produces countless winnings.

The numbers can be switched between 4 and 8. Once it is switched the 188 will automatically become 144, which will definitely produce the winning number 411.

The hidden beauty here is that other winning numbers in the trend are altered accordingly. Which is why the "Lottery Maestro Signature" is the system that has not failed and will never fail in lottery betting.

The previous group after switching between the 4 and 8 will become the new group below:

3	8	2
8	0	1
1	4	4
4	1	0
3	2	8

We are now clearly closer to producing the winning Pick 3 number 411. The process will be to convert the numbers into Shadow. Always remember that you will do the same to the rest of the group. All winning trends exhibit similar behavior and as such any conversion done to one set must be done to the rest of the group for the trend to continue.

After converting the entire group into Shadow, the above group will become the new one below:

0	2	8
2	3	4
4	1	1
1	4	3
0	8	2

We have now recreated the 411 that Maryland played on 5/1/19. The next winning Pick 3 number that played in the evening of that same day was 553.

I will begin the process of recreating the 553 winning number by looking at how I can first and foremost produce any number or numbers remotely close to 553. The first thing I will do is to convert the above group into Counterpart which changes the entire group into:

5	7	3
7	8	9
9	6	6
6	9	8
5	3	7

At this juncture, even if you decided to keep playing the rest of the group, you will notice that 789, 966 and 698 played in that same month of May 2019. We are now talking about capturing that many winning Pick 3 numbers in the same month based on the very 1st number that played for the month. That is unheard of. You can only be this good when you master how to use the "Lottery Maestro Signature." This was possible because we did not take the bet of working out the winning Pick 3 number 411 from 401 rather than going through the least expected route 188 that exposed the winning trend. The other winning number 573 in the group played one month earlier on 4/14/19 which would serve as confirmation that this trend is on the money.

Now back to recreating the 553 that played on that same day.

We can easily recreate the winning number 553 by changing the numbers in the middle digit into Shadow which would change the "7" of 573 into "5," thereby making it 553. Once you apply the Shadow, the entire group above will become:

5	5	3
7	2	9
9	9	6
6	6	8
5	0	7

The "Lottery Maestro Signature" understands that the readily available winning number is not necessarily where the gold is. Granted, we exposed the winning Maryland Pick 3 number 553, but that's not enough. If you look at the trend before last, you will notice that it has the winning Pick 3 number 573 from which 553 is derived but also has the number 537.

What makes you think that they won't form the 553 from the 537 instead of 573 or from any other winning Pick 3 number in the group?

Remember that the real winners are hidden in the least expected places which is why we are not going to settle for the initial winning Pick 3 number 553.

We can recreate the winning Maryland Pick 3 number 553 of 5/1/19 from 668 by switching between the numbers 6 and 7 i.e. replacing the 7's with 6 and the 6 with 7.

Once they are switched, the previous group will become:

5	5	3
6	2	9
9	9	7
7	7	8
5	0	6

You will notice that the earlier 573 and 537 that are similar and can produce the same winnings is now diverging into two completely different winning Pick 3 numbers. If 573 is 553, then 537 will definitely become 506 in which the later are not similar and cannot produce winnings alike.

What I am saying here is that the entire trend is shifting and exposing the hidden trend once you apply the "Lottery Maestro Signature."

We will, in essence, reproduce the winning Pick 3 number 553 from a different space. The winning Pick 3 number 778 can be converted by using Shadow for the 1st two digits thereby changing the "77" of 778 into "55" and then apply Counterpart afterwards to the "8" of 778 to produce 3.

Once you apply the conversions, the entire group above will become:

7	7	8
9	8	4
6	6	2
5	5	3
7	3	1

As you can see, I did not take the bet of settling with the initial winning Pick 3 number 553. I chose to recreate it from a different Pick 3 number among the group which ultimately paid off by exposing other winning numbers for the month like 553, 266 and 984 right away. You might be tempted to believe that the 731 and 778 did not play.

As I stated earlier, "Lottery Maestro Signature" does not fail. If you convert the last digits of 778 and 731 into Shadow and Counterpart, the two winning numbers will change into 777 and 739 both of which played in the same month of May 2019.

You can also switch more than one number simultaneously. This exposes the current trend when you line them up side by side.

We are looking for 100 dedicated lottery players to sign up for the upcoming "26 Master Series." This will be a bi-weekly series that analyzes lottery results from different states and predicts the numbers that will play for the next two weeks. It is also meant to be a learning tool with the goal of making the participants the best when it comes to betting on lottery. The series will be between $1 and $3 to cover the costs. Our only interest is in helping you become one of the best lottery players in the world. If you're interested in becoming one of the first 100 people to participate in our "26 Master Series," send a one line message to include you in this program.

All of these messages should be sent to:

LotteryMaestro@yahoo.com

Master Lottery Maestro, and break the lottery bank!

Always seek the TRUTH relentlessly, Embrace the TRUTH warmly, Live the TRUTH genuinely and Disseminate the TRUTH lavishly.

With Love

Eze Ugbor